Georg Ossian Sars

Crustacea Caspia

Contributions to the knowledge of the carcinological fauna of the Caspian Sea

Georg Ossian Sars

Crustacea Caspia
Contributions to the knowledge of the carcinological fauna of the Caspian Sea

ISBN/EAN: 9783337256555

Printed in Europe, USA, Canada, Australia, Japan

Cover: Foto ©berggeist007 / pixelio.de

More available books at **www.hansebooks.com**

ИЗВѢСТІЯ
ИМПЕРАТОРСКОЙ АКАДЕМІИ НАУКЪ.

ТОМЪ I. № 2.

1894. ОКТЯБРЬ.

BULLETIN
DE
L'ACADÉMIE IMPÉRIALE DES SCIENCES
DE
ST.-PÉTERSBOURG.

Vᵉ SÉRIE. TOME I. № 2.

1894. OCTOBRE.

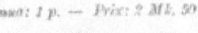

С.-ПЕТЕРБУРГЪ. 1894. ST.-PÉTERSBOURG.

Продается у коммиссіонеровъ Императорской Академіи Наукъ:
И. Глазунова, М. Эггерса и Комп. и К. Л. Риккера въ С.-Петербургѣ,
Н. Киммеля въ Ригѣ,
Фоссъ (Г. Гесселъ) въ Лейпцигѣ.

Commissionnaires de l'Académie Impériale des Sciences:
MM. J. Glasounof, Eggers & Cie. et C. Ricker à St.-Pétersbourg,
N. Kymmel à Riga,
Voss' Sortiment (G. Haessel) à Leipzig.

Цѣна: 1 р. — Prix: 2 Mk. 50 Pf.

Crustacea caspia.

Contributions to the knowledge of the Carcinological Fauna of the Caspian Sea.

By G. O. Sars,

Professor of Zoology at the University of Christiania, Norway.

Part III.

AMPHIPODA.

1-st Article.

Gammaridæ (part).

With 8 autographic plates.

(Lu le 11 mai 1894).

INTRODUCTION.

The Caspian Sea would seem truly to abound in Amphipoda. By the investigations of Dr. Grimm and Mr. Warpachowsky, a rather extensive material has now been brought together, the examination of which shows indeed the Amphipodous Fauna of that isolated basin to be both rich and diversified, comprising, as it does, numerous species belonging to several distinct families. As yet known, the following families are represented in the Caspian Sea: *Lysianassidae, Pontoporeiidae, Gammaridae, Corophiidae*. The 2 last-named families are represented both in the collection of Mr. Warpachowsky and that of Dr. Grimm, whereas only the latter collection contains forms belonging to the 2 first ones. Of the above mentioned 4 families, the *Lysianassidae* and *Corophiidae* are, as well known, exclusively marine in character, whereas the other 2 comprise, besides marine species, also some apparently genuine fresh-water forms. By far the most abundantly represented family is that of the *Gammaridae*, and of the genera comprised within it, the genus *Gammarus* has turned out contain much the greater part of the species. Some of the Gammaroid genera occurring in the Caspian Sea are very remarkable and rather unlike those represented in the Oceans. Especially is the generic form named by Dr. Grimm *Boeckia* highly distinguished by its most strange appearance.

As is the case with the *Mysidae* and *Cumacea*, much the greater part of the Caspian Amphipoda are, as yet known, restricted in their occurrence to that basin, only a few forms having been stated to be common also to the Black Sea. According to the investigations of Dr. Grimm, several forms descend to very considerable depths, and among them are some, which evidently show themselves to be of true arctic origin.

Our knowledge to the Caspian Amphipoda is still very imperfect, only scattered notes having hitherto been published about this part of the Fauna. It therefore cannot fail that a full account of the species occurring in that isolated basin may have a considerable interest. I give below a summary of the earlier publications referring to the Amphipodous Fauna of the Caspian Sea, as far as I have been enabled to state by looking over the literature accessible to me.

In his «Fauna caspio-caucasia»[1]), Mr. Eichwald mentions 2 species of *Gammarus* occurring in the Caspian Sea, and already noticed many years previously by Pallas[2]). The one of these species was considered by the latter author as identical with *G. pulex* Fabr., whereas the other was noticed as a new species and named *G. caspius*. Mr. Eichwald gives a short diagnosis of the latter form, and describes more at length another species from the Black Sea, *G. haemobaphes*, which he believes is the same as that noticed by Pallas as *G. pulex*. In the Catalogue of Amphipoda in the British Museum (1862), Sp. Bate describes and figures 2 species of *Gammarus*, *G. caspius* Brandt and *G. semicarinatus* n. sp., which both would seem to belong to the Caspian Fauna, though no exact locality was indicated for any of them. The last-named species is unquestionably, to judge from the figure, identical with *G. caspius* of Pallas as characterised by Eichwald, whereas the former is a very different species, perhaps that subsequently named by Dr. Grimm *G. aralo-caspius* (or *G. robustoides*). Sp. Bate refers for this species to Brandt's treatise in Middendorff's Sibirische Reise, but this must be an error, as no species of that name is mentioned in that work; and the locality (Asiatic Russia?) would seem to have merely been inserted because the specimen, from which the description and figure was taken, was presented to the Museum of the Jardin des Plantes by Professor Brandt. The most recent publication referring to the Amphipodous Fauna of the Caspian Sea is that given by Dr. Grimm in «Archiv für Naturgeschichte» for 1880[3]). In this very interesting treatise no less than 18 different species of Caspian Amphipoda are mentioned, collected by him from rather deep water in the southern and middle part of that Sea. But the species are only named, no descriptions whatever having

---

1) Nouv. Mém. de la Soc. Imp. des Naturalistes de Moscou, T. VII, 1842.
2) «Reise durch Russland I. 1801» (according to Eichwald).
3) «Beitrag zur Kenntniss einiger blinden Amphipoden des Kaspisces.»

been given of any of them. It is only some few points in their organisation (especially the more or less development of the visual organs), which have been treated of in detail, and this treatise is thus quite insufficient for recognizing any of the species named.

As, however, the collection of Dr. Grimm has kindly been placed in my hands for examination, and some of the specimens contained in it are labelled with the names given to them, I have been enabled to identify several of the forms collected by Mr. Warpachowsky in the northern part of the Caspian Sea with species detected at an earlier date by Dr. Grimm, and I have endeavoured in every possible case to retain for the species the names originally given to them by that distinguished naturalist, though in some cases it has been necessary to make a slight change with the names, partly because they have been preoccupied in Zoology, and partly because they have been less correctly formed. It may be noticed that some of the species named in his above-cited treatise (for instance *Gammarus Grigorkowii*, *G. coronifera*, *G. thaumops*) do not seem to be contained in the collection sent to me, and that some others are only represented by apparently quite immature specimens, which hardly suffice for recognizing the species. Moreover some of the specimens have been dried up at an earlier date by the evaporation of the spirit, and on this cause deformed, so as to be only with great difficulty examined. The greater part of the specimens are, however, still in a very good state of preservation, and will suffice for a full examination of the species.

The description of the new species contained in the collection of Dr. Grimm must be suspended for some time, as it has been destined, that the results of the investigations of that naturalist should be published in a separate work. It is therefore only the collection made by Mr. Warpachowsky in the northern part of the Caspian Sea, that will be the object of the present treatise. I have however been authorized by the Academy to refer to the collection of Dr. Grimm, as regards the horizontal and vertical distribution of the species here described.

The collection of Warpachowsky contains no less than 25 different species, and as some of these species are very nearly allied, and moreover the sexual differences often rather pronounced, it has appeared to me desirable, that each species should be described and figured in detail, and that also good and sufficiently large habitus-figures should be given of both sexes. On this cause it has been necessary to divide my treatise on that part of the Fauna into several articles, each accompanied by 8 plates. The present 1st article will give full descriptions and figures of 7 species belonging to 4 different Gammaroid genera, viz., *Boeckia*, *Gmelina*, *Amathillina* and

Gammarus. In the next article, to be shortly published, the remaining species of *Gammarus* will be described, and in a 3rd article some other Gammaroid genera will be treated off, as also the rather numerous species of *Corophium* occurring in the Caspian Sea.

The figures are, as in my two former papers reproduced by the autographic methode, and particular care has been applied in making them as correct and instructive as possible.

Fam. GAMMARIDÆ.

Gen. 1. **Boeckia**, Grimm (not Malm).

Generic characteristic. — Body very robust, with greatly incrusted integuments, and having the metasome and urosome poorly developed. Segments of mesosome produced laterally to extant spiniform processes, that of the 5th segment being particularly strong and mucroniform. Cephalon produced in front to a distinct rostrum, and having on each side a greatly prominent spiniform projection. Anterior pairs of coxal plates rather deep; 4th pair but little broader than the preceding pairs, and very slightly emarginated posteriorly. Eyes distinct, placed on the lateral faces of the cephalon. Superior antennæ longer than the inferior and having the accessory appendage obsolete. Oral parts normal. Gnathopoda comparatively small, subcheliform, and but little different; those in male somewhat stronger built than in female, with the propodos broader. Pereiopoda rather elongated and nearly equal in length, basal joint of last pair broader and more laminar than that of the 2 preceding pairs. Branchial lamellæ large, subpedunculated; incubatory lamellæ well developed. Uropoda very unequal in size, the last pair being rather small, not nearly reaching beyond the others and having the inner ramus extremely minute, scale-like, the outer linear and without any terminal joint. Telson very small, unarmed, and slightly cleft at the tip.

Remarks. — The name *Boeckia*, it is true, has been long ago appropriated in Zoology, having even been proposed at different times by 2 different authors, viz., by Malm for a genus of Amphipoda, and by Mr. Geo. Thomson for a fresh-water Copepod. But in both instances the name has been withdrawn, that of Malm being synonymous with *Leptocheirus* of Zaddach, and that of Thomson having been changed by Mrss. Guérin and Richard to *Boeckella*. It seems to me therefore that there cannot be any objection in using this name now in a new sense, and in every case I find it unreasonable that the name of such a diligent investigator as the late Dr. Boeck should not be justly associated with the order of Crustacea that constituted his special study, merely because some authors erroneously applied his name for the establishment of spurious genera.

The systematic position of this remarkable genus would seem, at the first sight, to be somewhat doubtful. In the robust form of the body, the poor development of the metasome and urosome, as also in the comparatively short caudal appendages, it rather much reminds of the *Orchestiidae*. But the oral parts are constructed upon the very same type as in the true *Gammaridae*, and the branchial lamellæ exhibit the same characteristic pedunculated appearance as in this family. Moreover the superior antennæ are considerably longer than the inferior, differing, however, very markedly from those in the other Gammaridæ in the want of a true accessory appendage. Notwithstanding this and other divergences from the Gammaroid type, I am inclined to believe, that this genus ought more properly to be placed within the *Gammaridae*, though constituting a rather anomalous membre of the family.

Besides the species described below, Dr. Grimm has distinguished 2 other species of this genus under the names *B. nasuta* and *B. hystrix*. Both these species are, however, founded upon quite immature specimens, the first-named agreeing exactly with young specimens of *B. spinosa*, as figured Pl. II. fig. 10; and the other only differing in the dorsal prominences of the segments being somewhat stronger and elevated to acutely triangular projections. In my opinion both these supposed species ought to be withdrawn, the genus being at present only represented by a single species.

1. Boeckia spinosa, Grimm.

(Pl. I and II).

Specific Characteristic. — Body in female extremely stout and very tumid, in male somewhat more slender and less broad; back obtusely carinated throughout, with the segments slightly projecting dorsally; mesosome having on each side, at the junction of the coxal plates, a row of spiniform processes, those of 5th segment being very large and terminating in a sharp point, the others comparatively small and obtuse at the tip; segments of metasome each with a pair of subdorsal, upturned processes, which however in the 1st segment are rather small and tuberculiform; 1st segment of urosome overlapping dorsally the succeeding ones and terminating in a rather large, hooked, median projection, having besides, as the segments of metasome, a pair of upturned subdorsal processes. Cephalon slightly keeled dorsally, rostrum horizontally projected and triangularly pointed, lateral projections longer than the rostrum, and diverging to each side nearly at a right angle. Anterior pairs of coxal plates much deeper than the corresponding segments, and having the distal edge densely setiferous; 1st pair somewhat narrower than the succeeding pairs and slightly curved; 4th pair not fully as deep as the preceeding pairs, and having the distal edge obliquely truncated; the 3 posterior pairs comparatively small and of normal appearance. Last pair of epimeral plates of metasome obtusangular. Eyes small, rounded, with dark pigment. Superior antennæ somewhat exceeding in length $\frac{1}{3}$ of the body, and rather densely setiferous, 1st joint of the peduncle comparatively large, flagellum nearly twice the length of the peduncle, and composed of numerous articulations; accessory appendage only represented by an extremely small nodule carrying 2 delicate bristles. Inferior antennæ but little more than half the length of the superior, and of normal structure. Gnathopoda in female rather feeble, subequal, propodos in both pairs about the length of the carpus and scarcely broader, palm well defined, being in the anterior ones more oblique than in the posterior; those in male somewhat stronger, with the propodos considerably expanded, forming below a rounded spiniferous lobe defining the deeply concaved palm, dactylus strong and curved. Pereiopoda rather slender, and having their outer part edged with numerous fascicles of bristles, basal joint of antepenultimate and penultimate pairs comparatively narrow and tapering distally, that of last pair considerably more expanded, with the greatest breadth below the middle. The 2 anterior pairs of uropoda having the rami subequal and falciform in shape; last pair much shorter than the former, with the outer ramus somewhat longer than the basal part, and provided with a few fascicles of small bristles. Telson extremely small, scarcely half as long as it is broad at the base, outer part narrowed and having in the middle a short cleft. Length of adult female 20 mm., of male 25 mm.

Remarks. — This remarkable Amphipod cannot be confounded with any other form, exhibiting, as it does, a most peculiar appearance by the extremely stout and compact body and its strange spinous armature. It may however be observed, that some of the species of *Allorchestes* (or *Hyallella*) found in the Titicaca Sea and described by Mr. W. Faxon, exhibit a somewhat analogous armament of the body. Especially is this the case with the species named *Allorchestes armata*. But here the lateral spines are not formed by the segments themselves, but by the greatly extant coxal plates, the most prominent lateral spines being represented by the 4th pair of coxal plates. It is evident that this peculiar armature of the body, occurring in a similar mode in so widely different forms, must have some significance for the animal. I believe that these laterally projecting acute spines may serve as a means of defence, whereby the animal, which apparently is far less active than the other Gammaridæ, becomes partly secured against the attack of fishes and other enemies. This may also apply to the above mentioned species of *Allorchestes*, which do not seem to be very habile swimmers.

Description of the female.

(See Pl. I).

The length of the body in adult ovigerous specimens measures, when fully extended, about 20 mm., and this Amphipod attains thus a rather large size.

The form of the body (see figs. 1 and 2) is extremely stout and compact, more so than in any other known Gammarid, and all the integuments are very hard and highly incrusted. In alcoholic specimens the body is generally found to exhibit a strong curvature, the posterior part being folded in beneath the anterior, and the head curved downwards. In this state it looks like an irregular ball, from the centre of which projets on each side the large mucroniform spine of the 5th segment. When fully extended, the back remains still somewhat curved (see fig. 1), though the mutual longitudinal relation of the several body-divisions now may easily be determined. It is found that the mesosome occupies much the greater part of the body, the metasome and urosome being comparatively poorly developed and combined scarcely longer than the former division. All the segments of the body appear very sharply defined, and those of the mesosome are particularly broad and subfornicate in shape, being produced on each side, just above the junction of the coxal plates, to rounded prominences, each tipped by a laterally projecting spiniform process. The 4 anterior and 2 posterior pairs of those processes are comparatively short and obtuse at the tip, whereas those of the 5th pair are very large and prominent, mucroniform, and gradually tapering to a very acute point. Along the back both the mesosome and metasome exhibit a distinct, though somewhat obtuse keel, which in each segment is elevated to a rounded dorsal prominence, those of the segments of the metasome being somewhat more compressed and sublaminar. In each of the latter segments occur near the dorsal face a pair of upturned digitiform processes, which however on the 1st segment are generally very small and merely tuberculiform. The 1st segment of the urosome (see fig. 5) is comparatively large and of a somewhat trigonal form, being produced at the end dorsally to a rather prominent and somewhat hooked projection, fully overlapping the 2 succeeding very short segments, and even reaching somewhat beyond the tip of the last pair of uropoda. At the base of this projection occur a pair of subdorsal digitiform processes of a similar appearance to those found in the posterior segments of the metasome.

The cephalon about equals in length the first 2 segments of mesosome combined. It exhibits dorsally a low keel, and is produced in front to a somewhat flattened, horizontally projected rostrum of an acute triangular form, and reaching nearly to the end of the basal joint of the superior antennæ. The lateral faces of the cephalon are evenly convex in their upper part, but inferiorly they jut out on each side to a remarkable spiniform process extending laterally nearly at a right angle to the longitudinal axis. These processes are considerably longer than the rostrum and terminate each in a sharp point.

The 4 anterior pairs of coxal plates, extending nearly vertically downwards, are rather large, being almost twice as deep as the corresponding segments. They are all densely fringed on the distal edge with delicate bristles, and, when the body is curved in the manner usually found in alcoholic specimens, completely overlap each other with their anterior edges, so as to form together on each side a continuous wall, inside which the oral parts, the gnathopoda, and partly also the 2 anterior pairs of pereiopoda may be wholly concealed. When the body is fully extended (see fig. 1), these coxal plates become somewhat separated in their outer part, still forming in their upper part a continuous wall. The 1st pair of coxal plates (see fig. 15) are somewhat narrower than the succeeding ones and slightly curved, with the anterior edge concave, and the outer part somewhat expanded, forming in front a narrowly rounded lobe, which, when the animal curves itself, is received just beneath the lateral process of the cephalon. The 2 succeeding pairs of coxal plates (see fig. 16) are nearly of equal size and oblong quadrangular in form, with the anterior corner somewhat more projecting than the posterior. The 4th pair (see also Pl. II, fig. 4) are not fully as deep as the 2 preceding pairs and but little broader. They exhibit a rather different form, being obliquely truncated at the end, with the posterior edge slightly emarginated in its upper part, and projecting below the emargination as an obtuse angle.

The 3 posterior pairs of coxal plates are much smaller than the anterior, and successively decrease in size. The 5th pair are scarcely half as deep as the 4th, and, as usual, divided into 2 rounded lobes, the anterior of which is somewhat deeper than the posterior. The 2 last pairs are transversely quadrangular in form.

The epimeral plates of the mesosome are not very large; those of the 2 anterior segments are rounded, those of the last segment obtus-angular.

In a dorsal view (fig. 2) the body appears very tumid and of a somewhat fusiform shape, the greatest breadth, which is fully as great as the height (including the coxal plates) and about equals $1/3$ of the length, occurring about in the middle, whence the body gradually tapers both anteriorly and posteriorly. The lateral spines become, in this view of the animal, very conspicuous, projecting, as they do, from each side of the mesosome. The extent between the tips of the large mucroniform processes of the 5th segment considerably exceeds half the length of the whole body.

The eyes (see figs. 1 and 2) are placed on the lateral faces of the cephalon, at some distance from the anterior edge and somewhat nearer the dorsal than the ventral side. They are comparatively small and of rounded form, with very dark pigment.

The superior antennæ (fig. 3) somewhat exceed in length $\frac{1}{3}$ of the body and are rather slender, being densely supplied with delicate bristles on both edges. They are very flexible and generally so much recurved, as to be nearly completely hidden between the lateral processes of the head and the coxal plates. Of the 3 joints of the peduncle the 1st is much the largest, equalling in length the other 2 combined and being much thicker. The last 2 joints of the peduncle are nearly of equal length, but the last is somewhat narrower than the 2nd.

The flagellum is nearly twice as long as the peduncle, and composed of numerous short setiferous articulations, their number amounting to about 25 in all. The accessory appendage seems at the first sight to be entirely wanting. On a closer examination, however, an extremely small nodule is found in the place, where in other Gammaridæ this appendage occurs. This nodule is distinctly defined from the last peduncular joint, and carries on the tip 2 delicate bristles.

The inferior antennæ (fig. 4) are much shorter than the superior, but little exceeding half their length, and, as the latter, are rather densely setiferous and generally strongly recurved. In every case their basal part remains quite hidden by the lateral processes of the head, and can only be examined by dissection. They are on the whole quite normally constructed, exhibiting a comparatively large globular basal joint, followed by a very short joint, from which inside the olfactory spine issues. The 3rd joint is likewise rather short but comparatively thick, whereas the 2 remaining joints of the peduncle are much more elongated, the penultimate one being the larger. The flagellum is about same length as the last 2 peduncular joints combined, and composed of 9 articulations.

The buccal area is not very much prominent, and scarcely visible in a lateral view of the animal, it being almost completely hidden between the 1st pair of coxal plates. The several oral parts composing it, are on the whole of a quite normal structure, agreeing with that generally found in the typical Gammaridæ.

The anterior lip (fig. 8) is of a rounded form, and somewhat narrowed in its outer part, with the tip scarcely emarginated and finely ciliated at the edge.

The posterior lip (fig. 9) is comparatively large and of the usual submembranaceous consistence. The lateral lobes are rather broad and, as usual, ciliated at the tip and the inner edge, whereas they outside project as an obtusely conical lappet. There is only a very slight rudiment of inner lobes.

The mandibles (figs. 10 and 11) are strongly built, with the molar expansion well developed and the cutting edge divided, as usual, into 2 superposed, dentated plates, somewhat differently shaped in the 2 mandibles. Between the cutting edge and the molar expansion occurs the usual series of curved, finely ciliated spines. The mandibular palp (see fig. 10) is of moderate size, being scarcely longer than the mandible itself. Its terminal joint is about as long as the 2nd, somewhat compressed, and gradually tapering distally. It carries on the inner edge a dense series of comparatively short, ciliated spinules, and has besides on the tip and the outer edge several slender bristles.

The 1st pair of maxillæ (fig. 12) exhibit the normal structure. The masticatory lobe is moderately strong, and armed on the truncated tip with a rather great number of partly denticulated spines arranged in a double row. The basal lobe is oval in form, and carries along the inner edge a row of about 10 ciliated setæ. The palp is, as in most other Gammaridæ, somewhat differently developed in the 2 maxillæ, its terminal joint being on the right maxilla very much expanded and having the distal edge divided into a number of coarse denticles, whereas on the left maxilla this joint is much narrower and provided at the tip with a few slender spines.

The 2nd pair of maxillæ (fig. 13) have the outer lobe a little larger than the inner, both being oblong oval in form and carrying at the tip a number of delicate, curved bristles. The inner lobe, moreover, is provided with about 6 ciliated setæ arranged in a somewhat oblique series on its lower face.

The maxillipeds (fig. 14) are, as usual, quite fused together at the base, springing off from a common basal part composed of 2 somewhat flattened segments. The basal lobes, springing off from the 2nd segment of the basal part and lying in close juxtaposition, are of moderate size and subquadrangular in shape. They carry at the tip a number of delicate curved bristles, between which there occur a few short denticles, and have the inner edge fringed with a series of ciliated setæ. The masticatory lobes are somewhat larger than the basal ones, and are armed along their inner edge with a series of flattened spines increasing in size towards the tip, where they successively become transformed to strong curved setæ. The palp is well developed, subpediform, and composed of the usual 3 joints, the outer 2 of which form together a more or less pronounced geniculate bend. The last joint is somewhat expanded in its outer part, which is densely setous, and carries at the tip a claw-like movable spine (the dactylus).

The gnathopoda (figs. 15 and 16) are comparatively small and most frequently so closely applied against the buccal area, as to be quite hidden between the 2 anterior pairs of coxal plates, to the inner face of which they are articulated. They are nearly alike both in size and structure, both pairs being densely setous and exhibiting a more or less pronounced sigmoid curve. The basal joint is somewhat more elongated in the posterior ones (fig. 16) than in the anterior, whereas the 3 succeeding joints are exactly alike in both pairs, the carpus being about the length of the 2 preceding joints combined and forming below a slight setous expansion. The propodos is about as long as the carpus and scarcely broader, exhibiting in both pairs a distinct subcheliform structure. Its shape is a little different in the 2 pairs, the palm being in the anterior ones (fig. 15) somewhat oblique, whereas in the posterior ones (fig. 16) it is nearly transverse. The dactylus is not very strong and of the length of the palm.

The pereiopoda (see fig. 1, comp. also Pl. II, figs. 4—7) are rather much elongated and but little different in length, all being fringed on both edges with numerous fascicles of short bristles, and having the dactylus rather slender. The 2 anterior pairs are, as usual, generally turned anteriorly, whereas the 3 posterior pairs are more or less strongly reflexed; in the former the basal joint is comparatively narrow, in the latter more lamellar in character. Of the several joints composing these limbs, the basal one is in all much the largest and the ischial joint the shortest, the 3 succeeding ones being nearly of equal length. The antepenultimate and penultimate pairs are somewhat longer than the others, and have the basal joint but little expanded and gradually tapering distally. The last pair (comp. Pl. II, fig. 7) are a little shorter than the 2 preceding pairs, and differ considerably in the form of the basal joint, which is much broader and considerably expanded in its distal part, the posterior edge being strongly curved below the middle and fringed throughout with short bristles.

The branchial lamellæ, present at the base of all the legs, except the anterior gnathopoda, are well developed, though, as usual, considerably diminishing in size posteriorly. The anterior pairs (see Pl. I, fig. 16) are rather large and broad, subtriangular in form, and attached by a short but well-marked peduncle inside the coxal plates, at some distance from the insertion of the basal joint of the corresponding leg. Close to them, and somewhat more inside issue the incubatory lamellæ, forming together the marsupial pouch. They are likewise rather large and fringed with long setæ.

The 3 pairs of pleopoda exhibit quite a normal appearance.

The uropoda are very unequal in size (see Pl. I, fig. 5), the 1st pair being much the largest and, as the succeeding pair (comp. Pl. II, fig. 8), having the rami subequal and somewhat falciform in shape, both gradually tapering distally, with a single small apical denticle and another about in the middle of the upper edge. The last pair (Pl. I, fig. 7) are extremely small, not at all reaching beyond the others, and are also rather different in structure. They consist each of a short and thick basal part, to the end of which are attached 2 very unequal rami. The outer one is a little longer than the basal part and of a narrow linear form, with 3 fascicles of small bristles along one of the edges and a similar one at the tip. The inner ramus is very minute and scale-like, with a single small spine at the tip.

The telson (fig. 6) is extremely small, and not easy to examine in the uninjured animal, it being completely overlapped by the hooked dorsal projection of the 1st segment of the urosome. When isolated by dissection, it shows itself to be of a broadly triangular form, with the length not nearly attaining half the breadth, and the tip cleft by a short and narrow incision. On the dorsal side of each of the narrowly rounded terminal lobes occurs a very small spinule; otherwise the telson is quite unarmed.

The adult male (see Pl. II) attains a still larger size than the female, the length of the body, when fully extended, amounting to no less than 25 mm. In general appearance it does not differ much from the female, except by the body being considerably less tumid. In a dorsal view of the animal (Pl. II, fig. 1) the sex may therefore at once be determined. Of the several appendages it is chiefly the gnathopoda, which distinguish themselves by a much stronger build than in the female (see figs. 2 and 3). Especially appears the propodos in both pairs much larger and considerably expanded, forming below a broadly rounded lobe defining the palm inferiorly and armed with a number of strong anteriorly curving spines. The palm is deeply concave, and the strongly curved dactylus impinges, when closed, with the tip somewhat inside the inferior expansion of the propodos. The pereiopoda are on the whole of the same structure as in the female, though being perhaps a little more elongated and having the basal joint of last pair somewhat less expanded. Finally, the outer ramus of the last pair of uropoda (fig. 9) appears a little longer and is provided on the inner edge with several slender bristles not found in the female. Of course no incubatory lamellæ are present in male specimens; but the branchial lamellæ (see figs. 3 and 5) exhibit the very same appearance as in the female.

Very young specimens (fig. 10), of a length of about 6 mm., differ from the adult in all the processes of the body (also those of the cephalon and of the 5th segment of mesosome) being digitiform in shape, terminating with an obtuse point, and moreover in the dorsal prominences of the segments being more strongly elevated, giving the back a serrated appearance. In all these particulars they exactly agree with the form named by Dr. Grimm *B. nasuta*.

Colour. — In none of the specimens examined any trace of pigmentary ornament could be detected, all exhibiting a uniform whitish colour. In the living state, however, the animal may most probably have shown some characteristic colouring.

Occurrence. — This remarkable form was met with by Mr. Warpachowsky in 4 different Stations of the North Caspian Sea, one of which (St. 7) was located at the mouth of the Bai Agrachansky, 2 others (St. 58, 59) between the Tschistyi-Bank and the mouth of the Wolga, and the 4th (St. 61) far North, at some distance outside the Bai Bogutyi Kultuk. In one of the Stations (58) several specimens, both males and females, were secured; in the other 3 Stations only solitary specimens occurred.

In the collection of Dr. Grimm 2 adult male specimens of this form are contained, found in 2 different Stations, the one located off the promontory Schachowa Kosa, the other at some distance South of the peninsula Mangyschlak, the depth being in the former 7 fms., in the latter 90—100 fms. Besides 2 immature specimens (= *B. nasuta* Gr.) were collected in the last-named Station, and another, likewise immature specimen (= *B. hystrix* Gr.) was procured in the southern part of the Caspian Sea, from the very considerable depth of 150 fms.

Out of the Caspian Sea this form has not yet been recorded.

Gen. 2. **Gmelina**, Grimm, MS.

Generic Characteristic. — Body slender and compressed, with rather strongly incrusted integuments, and having the segments sharply defined, partly also produced to conspicuous projections. Metasome and urosome well developed. Cephalon but slightly projecting in front, lateral lobes comparatively small, postantennal corners well defined and rather deep. Anterior pairs of coxal plates of moderate size and larger in the female than in male; 4th pair not very much expanded, and but slightly emarginated posteriorly in their upper part. Eyes well developed and more or less protuberant, being placed near the anterior edges of the cephalon. Antennæ not very much elongated, and nearly equal in length, the superior ones with a very small, uniarticulate accessory appendage. Oral parts normal. Gnathopoda in female rather feeble, though distinctly subcheliform, in male much more strongly developed and subequal, propodos very large and gradually widening distally. Pereiopoda not very much elongated, the 3 posterior pairs successively increasing in length; last pair having the basal joint somewhat larger and more lamellar than in the 2 preceding pairs. Last pair of uropoda more or less projecting beyond the others, and having the outer ramus well developed and more or less pronouncedly foliaceous in structure, inner ramus small, squamiform. Telson deeply cleft.

Remarks. — The present genus is very nearly allied to the genus *Pallasiella* G. O. Sars (*Pallasia* Sp. Bate), and indeed at first, before the collection of Dr. Grimm was come in my hands, I referred the 2 species described below to that genus. There is, however, perhaps some raison for supporting the new genus proposed by Dr. Grimm, since the said species exhibit some apparently essential points of difference from the type of the genus *Pallasiella*, for instance the much feebler structure of the gnathopoda, and the fact, that the telson is deeply cleft, not as in the latter genus only emarginated at the tip.

Besides the form upon which Dr. Grimm founded his genus, another very distinct species is contained in the collection of Mr. Warpachowsky, and this species has recently also been recorded from the Azow Sea by Mr. Sowinsky.

2. Gmelina costata, Grimm, MS.

(Pl. III).

Specific Characteristic. — Body extremely slender and compressed, especially in the male, with the lateral parts of the segments of mesosome somewhat exstant; back keeled throughout, the keel being elevated in the posterior segments of mesosome and those of metasome to conspicuous dorsal projections. Urosome unusually elongated, and having each of the 2 anterior segments produced dorsally to a small dentiform projection. Cephalon with the lateral faces quite smooth, rostral projection well-marked, lateral lobes but very little projecting and broadly truncated at the tip. Anterior pairs of coxal plates in female much deeper than the corresponding segments, in male considerably smaller; 4th pair but little broader than the preceding pair. Second pair of epimeral plates of metasome rather deep and acutely produced, last pair somewhat smaller and less produced at the lateral corners. Eyes of moderate size and but slightly protuberant, oval reniform, with dark pigment. Superior antennae a little longer than the inferior, but scarcely exceeding in length ⅓ of the body, joints of the peduncle successively decreasing in size, flagellum but little longer than the peduncle, accessory appendage not attaining the length of the 1st articulation of the flagellum. Gnathopoda in female somewhat unequal, the posterior ones being a little more slender, and having the propodos narrower; those in male much larger, with the propodos oblong oval in form, palm concave and defined below by a nearly rectangular projection armed with 2 strong spines, dactylus very strong and curved. The 3 posterior pairs of pereiopoda comparatively short and stout, and having their outer part edged with scattered fascicles of spines and delicate bristles, basal joint of the 2 anterior pairs rather small and tapering distally, that of last pair oblong quadrangular in shape. Last pair of uropoda rather fully developed and projecting far beyond the others, outer ramus very large, pronouncedly foliaceous and edged with slender spines and delicate bristles, tip blunt, with a very small terminal joint. Telson rather large, projecting beyond the basal part of the last pair of uropoda, cleft extending nearly to the base, terminal lobes obtusely pointed, and carrying each a single apical spine and a few delicate bristles. Length of adult female 12 mm., of male 16 mm.

Remarks. — The present form is at once recognized by its extremely slender and narrow body, on which cause I at first noted it under the provisional name *Palasiella macera*. The pronounced foliaceous character of the outer ramus of the last pair of uropoda may also serve for distinguishing this form from most of the other Caspian Amphipoda. It is the form upon which Dr. Grimm founded his genus *Gmelina*.

Description of the female.

Fully adult, ovigerous specimens attain a length of about 12 mm.

The general form of the body (see fig. 1) is very slender and highly compressed, the metasome and urosome being both well developed and combined about the length of the mesosome. The integuments are highly incrusted, exhibiting in some places, for instance in the anterior part of the coxal plates (see fig. 11) conspicuous rounded indurations. All the segments are very sharply marked off from each other, whereby the outer contours of the body acquire, both in the lateral and dorsal view of the animal, a somewhat rugged appearance. The segments of the mesosome have their lateral parts slightly prominent at the junction of the coxal plates, forming together an obtuse keel extending along each side of that division of the body. Another keel runs along the back, being anteriorly rather low, but gradually becoming more distinct backwards, and being at the same time successively elevated in the segments to more or less conspicuous dorsal projections. It is not easy to indicate with exactness where those projections take their begin, as they are only little by little growing out from the segments, but in the antepenultimate segment of the mesosome there is generally found a distinct approach to such a projection, and in the last segment, as also in those of the metasome, they are very conspicuous, being obtusely triangular in form and distinctly laminar. The urosome is unusually prolonged, nearly equalling in length the metasome, and has the 1st segment slightly keeled dorsally in its posterior part and produced at the end to a short acute projection; a similar, but much smaller, dorsal projection may also be observed in the succeeding segment, whereas the last segment is quite smooth above.

The cephalon (fig. 2) about equals in length the first 2 segments of mesosome combined, and is produced in front to a distinct, though not very large rostral projection. The lateral lobes are very slightly projecting and broadly truncated at the tip, being defined from the rather deep and acutangular postantennal corners by a slight emargination. The lateral faces of the cephalon are quite smooth, without any trace of a projection.

The 4 anterior pairs of coxal plates (see fig. 1) are rather large, being considerably deeper than the corresponding segments, and of an oblong quadrangular form, with only a few scattered hairs on the distal edge. The 1st pair (see fig. 11) are somewhat smaller than the succeeding ones, and very slightly expanded in their outer part. The 4th pair are but little broader than the preceeding pair, and exhibit posteriorly in their upper part a very slight emargination defined below by an obtuse angle.

The 3 posterior pairs of coxal plates are, as usual, much smaller than the anterior, and successively diminish in size. The 5th pair are but little broader than they are deeep, and have the anterior lobe somewhat more projecting than the posterior.

Of the epimeral plates of the metasome, the 1st pair are, as usual, the smallest and evenly rounded. The 2nd pair are considerably deeper and acutangular at the lateral corners; the last pair are of a more rounded form, though produced at the lateral corners to a short acute point.

The eyes (see fig. 2) are of moderate size and oval reniform in shape. They are but slightly protuberant and placed near the anterior edges of the cephalon. The pigment is dark.

The superior antennae (see fig. 1) scarcely exceed in length $\frac{1}{3}$ of the body, and are but sparingly supplied with small bristles. Of the joints of the peduncle the 1st is much the largest, being nearly as long as the other 2 combined. The last peduncular joint is considerably smaller than the 2nd. The flagellum is but little longer than the peduncle and composed of about

16 short articulations. The accessory appendage (see fig. 3) is distinctly defined, but rather small, and only composed of a single articulation carrying at the tip 3 slender bristles.

The inferior antennæ are a little shorter than the superior and, as the latter, but sparingly setiferous. Of the joints of the peduncle the penultimate one is the largest. The flagellum is about half the length of the peduncle and composed of 6 articulations.

The buccal area is somewhat projecting, though partly concealed by the 1st pair of coxal plates. The several oral parts composing it are on the whole quite normally constructed.

The anterior lip (fig. 4) exhibits the usual rounded form, and has in front an obtuse prominence.

The posterior lip (fig. 5) does not exhibit any trace of inner lobes. The lateral lobes are narrowly rounded in front, and project outside as an obtusely conical lappet.

The mandibles (figs. 6 and 7) are short and stout, and exhibit the usual armature of their masticatory part. The palp (see fig. 7) is rather slender, being considerably longer than the mandible itself, and has the last joint shorter than the 2nd.

The 1st pair of maxillæ (fig. 8) are comparatively large, with the masticatory lobe rather strongly developed and armed at the tip with coarse, denticulated spines. The basal lobe is subtriangular in form, and carries on the inner edge a row of about 8 setæ. The palp has the terminal joint on the left maxilla rather narrow, on the right, as usual, somewhat more expanded.

The 2nd pair of maxillæ (fig. 9) have the outer lobe considerably broader than the inner, exhibiting otherwise the usual structure.

The maxillipeds (fig. 10) in nearly all their details agree so closely with those in the preceding genus, that a detailed description of them is not needed.

The gnathopoda (figs. 11 and 12) are comparatively small and feeble in structure, though distinctly subcheliform and rather densely setous. They are a little unequal, the posterior ones being somewhat more slender than the anterior, and having the carpus larger. The propodos of the anterior pair (fig. 11) is oblong quadrangular in form and somewhat longer than the carpus, with the palm rather oblique; in the posterior pair (fig. 12) it equals in length the carpus and is somewhat narrower, with the palm nearly transverse.

The pereiopoda (see fig. 1) are comparatively short and stout, and rather unequal in length. The 2 anterior pairs are of same structure, though somewhat differing in length, the 1st pair being the longer.

The 3 posterior pairs successively increase in length, and have their outer part fringed with scattered fascicles of spines and delicate bristles, the dactylus being rather stout and curved, with a small denticle somewhat inside the tip. The antepenultimate pair are much shorter than any of the other pairs and, as the succeeding pair, have the basal joint comparatively small and narrowed distally. The last pair (fig. 13) differ from the preceding pairs in the much larger size of the basal joint, which is oblong quadrangular in form, with the posterior edge nearly straight and edged with scattered short hairs.

The uropoda are very unequal in size, the penultimate pair (fig. 14) being rather small, with the rami narrow linear and spinous only at the tip.

The last pair of uropoda (fig. 15) are of considerable size, projecting far beyond the others and nearly equalling in length the urosome. The basal part is short and thick, and the rami very unequal, the inner one being extremely small and scale-like, whereas the outer is very large and pronouncedly foliaceous in structure. It is nearly of equal breadth throughout and terminates with a blunted tip carrying an extremely minute terminal joint. The edges of the ramus are densely fringed with comparatively short, partly ciliated setæ, and are besides armed with fascicles of slender spines.

The telson (fig. 16) is comparatively rather fully developed, being considerably longer than it is broad at the base, and projecting beyond the basal part of the last pair of uropoda. It is divided by a deep cleft into two obtusely pointed lobes, which are finely ciliated on the outer edge and carrie each at the tip a single short spinule and a few delicate bristles.

The adult male (figs. 17, 18) grows to a considerably larger size than the female, reaching, when fully extended, a length of 16 mm. (excluding the last pair of uropoda).

The form of the body appears still more slender than in the female, and is also more compressed. In a dorsal view of the animal (fig. 18) the body therefore exhibits an extremely narrow, almost linear form. The sexual differences otherwise refer chiefly to the antennæ, the gnathopoda and the last pair of uropoda.

The antennæ (see fig. 17) appear somewhat more elongated than in the female, and also less unequal, the inferior ones being about same length as the superior. In both pairs, moreover, the flagella are composed of a greater number of articulations.

The gnathopoda (figs. 19 and 20) are very different from those in the female, being much more strongly built and nearly equal both in size and structure. In both pairs the propodos is very large, oblong oval, or rather somewhat clavate in form, gradually widening somewhat distally, with the palm distinctly concave, and defined below by a nearly rectangular projecting lobe armed with 2 strong spines, between which the strongly curved dactylus impinges, when closed.

The last pair of uropoda (fig. 22) are still larger than in the female, exceeding even considerably the whole urosome in length. This is chiefly caused by the fuller development of the outer ramus, the structure of which otherwise agrees with that in the female.

Colour. — All the specimens examined exhibited a uniform greyish white colour, without any conspicuous pigmentary marks; but this may most probably not have been the case in the living state of the animal.

Occurrence. — This form has been collected by Mr. Warpachowsky in 4 different Stations of the North Caspian Sea. Two of these (St. 16 and 52) were located off the island Podgornoj, another (St. 49) between the islands Kulaly and Morskay, the 4th (St. 58) at some distance north of the Tschistyi-Bank. In the latter Station only a single specimen was secured, in each of the others several specimens occurred.

Dr. Grimm collected the species at Baku, from the shores down to 6 fathoms, and moreover at the west coast of Sara among Zostera, and at Krasnowodsk in a depth of 20 fms.

Out of the Caspian Sea this species has not yet been recorded.

3. Gmelina Kusnezowi (Sowinsky).

(Pl. IV).

Gammarus Kusnezowi, Sowinsky, Les Crustacés de la mer d'Azow, p. 95, Pl. VIII.

Specific Characteristic. — Body rather slender and compressed, especially in the male, the back being, however, not carinated, but having a double series of tuberculiform projections, successively increasing in size, and assuming on the posterior segments of mesosome and those of metasome a mammilliform shape. Segments of mesosome (except the last 2) produced on each side, just above the junction of the coxal plates to very conspicuous, lateraly projecting rounded prominences. Segments of urosome smooth above, the last 2 having on each side dorsally 2 small spinules. Cephalon considerably attenuated in front and having on each side a conspicuous, umboniform prominence, rostral projection extremely small, lateral lobes narrowly rounded in front. Anterior pairs of coxal plates rather deep and of a similar shape to those in the preceding species: 5th pair somewhat oblique and much deeper anteriorly than posteriorly. The last 2 pairs of epimeral plates of metasome nearly rectangular. Eyes oval reniform and highly protuberant, being placed close to the anterior extremity of the cephalon. Antennae nearly equal-sized and rather short, scarcely exceeding in female ¼ of the length of the body. Gnathopoda nearly as in the preceding species, and exhibiting a similar difference in the two sexes. Pereiopoda likewise of a structure very similar to that in the said species, though being perhaps a little more slender. Last pair of uropoda not nearly so much elongated as in *G. costata*, the outer ramus being far less fully developed and also less pronouncedly foliaceous in character. Telson rather short, cleft narrow and extending nearly to the base, terminal lobes obtusely rounded and armed with several spines both at the tip and the outer edge. Length of adult female 14 mm., of male 18 mm.

Remarks. — There cannot be any doubt that the above-characterised form is that recently described by Mr. Sowinsky from the Asow Sea as *Gammarus Kusnezowi*. It is, however, certainly not a true *Gammarus*, but ought, in spite of the rather different armature of the body and the less fully developed last pair of uropoda, to be referred to the same genus as the preceding species, with which it agrees very closely in nearly all anatomical details. It is a very easily recognizable form, being highly distinguished by the peculiar subdorsal, mammilliform projections, on which cause I at first noted it under the provisional name of *Pallasiella mammillifera*.

Description of the female.

Adult ovigerous specimens attain, when fully extended, a length of about 14 mm.

The form of the body (see fig. 1) is rather slender and compressed, though perhaps not to such a degree as in the preceding species. As in the latter, all the integuments are highly incrusted, and the segments sharply marked off from each other.

The mutual longitudinal relation of the several body-divisions is about as in that species, except that the urosome is somewhat shorter. The body is generally more or less strongly curved, and has the back rounded off, not, as in the preceding species, carinated. On the other hand, there occurs along the back a double series of subdorsal prominences (one pair in each segment), which anteriorly are very low and tuberculiform, but farther back, on the last 2 segments of mesosome and those of metasome, assume a distinctly mammilliform shape, and, when the animal is viewed laterally, considerably project beyond the dorsale line. The lateral parts of the 5 anterior segments of the mesosome are, moreover, just above the junction of the coxal plates, produced to very conspicuous laterally projecting, tuberculiform prominences, best seen in a dorsal view of the animal (comp. fig. 13). The segments of the urosome are smooth above, without any projections, but, as in most species of the genus *Gammarus*, there occurs in the 2 posterior ones, on each side of the dorsal face, a fascicle of small spinules, their number being generally 2 in each fascicle.

The cephalon is somewhat shorter than the first 2 segments of the mesosome combined, and exhibits a rather irregular form. As seen laterally (fig. 2) it rapidly tapers anteriorly, being narrowly truncated at the tip, with the rostral projection extremely small and the lateral lobes narrowly rounded in front. The inferior edges of the cephalon between the latter and the postantennal corners are nearly straight and obliquely descending, and just above them issues from the lateral faces on each side a rather large umboniform prominence, best seen in the dorsal view of the animal (comp. fig. 13).

The coxal plates nearly agree in their shape with those in the preceding species, the 4 anterior pairs being rather large and considerably deeper than the corresponding segments. The 5th pair (see fig. 7) are somewhat oblique and much deeper in their anterior than posterior part.

The epimeral plates of the metasome are well developed, the 1st pair being, as usual, rounded, whereas the 2 succeeding pairs are nearly rectangular, with the lateral corners but slightly produced.

The eyes (see fig. 2), which are placed close to the extremity of the cephalon, are of oval reniform shape and remarkable by being so highly protuberant as nearly to exhibit a stalked appearance (comp. fig. 13). They have the visual elements well developed and the pigment of a very dark hue.

The superior antennæ (see fig. 1) are comparatively short, scarcely exceeding in length ¼ of the body, and are, as in the preceding species, but sparingly setiferous. The 1st joint of the peduncle is but little longer than the 2nd, and the 3rd only half the length of the latter. The flagellum does not attain the length of the peduncle, and is composed of about 15 short articulations. The accessory appendage (see fig. 3) is very small and exactly of same appearance as in the preceding species.

The inferior antennæ are about same length as the superior, and have the penultimate joint of the peduncle the largest. The flagellum is scarcely half as long as the peduncle and composed of 6 articulations.

The gnathopoda (figs. 4 and 5) exhibit a structure very similar to that in the preceding species, the posterior ones (fig. 5) being a little more elongated than the anterior and having the carpus somewhat larger. The propodos is in both pairs oblong oval in form, and but little broader than the carpus, with the palm somewhat oblique, though not nearly as long as the hind margin.

The 2 anterior pairs of pereiopoda (fig. 6) do not differ in any way from those in the preceding species, and also the posterior pairs (figs. 7 and 8) exhibit much the same structure, though being perhaps a little more slender and less coarsely spinous in their outer part. The basal joint of the last pair (fig. 8), as in *G. costata*, is considerably larger than that of the 2 preceding pairs, and exhibit a similar oblong quadrangular form.

The 2 anterior pairs of uropoda (fig. 9) are likewise of much the same structure as in that species.

The last pair of uropoda (fig. 10), on the other hand, do not nearly attain such a large size as in *G. costata*, though they somewhat project beyond the others. The rami are, as in that species, very unequal, the inner one being very small and scale-like, whereas the outer ramus is well developed and about twice as long as the basal part. This ramus does not, however, exhibit such a pronouncedly foliaceous character as in *G. costata*, being gradually narrowed distally and having the edges fringed with only a restricted number of slender spines and short bristles; at the tip occurs a distinct, though rather small terminal joint carrying several delicate bristles.

The telson (fig. 11) is comparatively small, not extending beyond the basal part of the last pair of uropoda. It is considerably broader than it is long, and divided by a narrow cleft into two obtusely rounded lobes, each armed with several spines (5—6 in number), 2 of which issue from the outer edge, the others close together from the tip.

The *adult male* (figs. 12 and 13) is considerably larger than the female, attaining a length of 18 mm.

The form of the body is about as in the female, though, as usual, somewhat more compressed, on which cause the body exhibits, in a dorsal view of the animal (fig. 13) a very narrow, nearly linear form.

The antennæ (see fig. 12) appear somewhat more elongated, though not nearly attaining ⅓ of the length of the body, and have a somewhat greater number of articulations in the flagella.

The gnathopoda (figs. 14, 15) are very strongly developed, and exhibit a structure closely agreeing with that in the preceding species, the propodos being in both pairs very large and of an oblong clavate form.

The last pair of uropoda (fig. 16) appear a little larger than in the female, with the outer ramus somewhat more elongated, but otherwise exhibit the very same structure.

This is also the case with the telson (fig. 17).

Colour. — In some of the specimens received short time after having been captured, there was still trace of a darkish pigment arranged in bands across the segments, similar to what occurs in the nearly allied Norwegian fresh-water Amphipod, *Pallasiella quadrispinosa*. Most of the specimens, however, exhibited a uniform greyish colour.

Occurrence. — This species also was collected by Mr. Warpachowsky in 4 different Stations. Three of these (St. 50, 58, 59) are located in the western part of the North Caspian Sea, whereas the 4th (St. 61) lies far north, at some distance outside the Bai Bogatyi Kultuk.

In the collection of Dr. Grimm this form is not represented.

Distribution. — The Azow Sea (Sowinsky).

Gen. 3. **Amathillina** [1]), Grimm.

Generic Characteristic. — Body comparatively robust, with the back to a more or less extent distinctly keeled, the keel being in all, or in some only of the segments elevated to compressed, posteriorly pointing projections. Urosome short and stout, without dorsal projections, but with fascicles of subdorsal spinules, as in the genus *Gammarus*. Integuments not very much incrusted. Cephalon with a small rostral projection, lateral lobes short and obtuse, postantennal corners well marked, lateral faces smooth. Anterior pairs of coxal plates of moderate size, 4th pair the largest and distinctly emarginated posteriorly in their upper part. Eyes well developed. Superior antennae slender and much longer than the inferior, with a well-developed accessory appendage. Oral parts normal. Gnathopoda in female rather feeble, though distinctly subcheliform; those in male very strongly built and nearly equal, exhibiting a structure similar to that in the male of the genus *Gmelina*. Pereiopoda of moderate length and edged in their outer part with fascicles of stiff bristles, dactylus in all strong and curved; last pair somewhat shorter than the penultimate one, and having the basal joint rather large and laminarly expanded. Last pair of uropoda comparatively small, scarcely reaching beyond the others, outer ramus sublinear, with scattered fascicles of spines, and having a distinct, narrow terminal joint, inner ramus small, squamiform. Telson short and broad, cleft to the base.

Remarks. — In the comparatively robust body, the back of which is to a more or less extent distinctly keeled and provided with lamellar dorsal projections, this genus somewhat reminds of the genus *Amathilla*. It differs, however, rather materially in the structure of the several appendages, and in this respect comes much nearer to the genus *Gammarus*, being chiefly distinguished from that genus by the poor development of the last pair of uropoda.

---

[1] Dr. Grimm spells the name *Amathillinella*, but this term cannot properly be accepted since it is a diminutive of *Amathillina*, a generic name which does not as yet exist. Probably Dr. Grimm had in view to form a diminutive of *Amathilla*, but this would correctly have been *Amathillella*, a name which would be inconvenient by its cacophony. Moreover *Amathilla* is itself a diminutive of *Amathia*, and to form a diminutive of a diminutive, would in every case seem to be objectionable.

In the collection of Mr. Warpachowsky 2 distinct, though nearly allied species are represented, one of which was named by Dr. Grimm, whereas the other is new to science. Besides Dr. Grimm has distinguished 2 other species as *A. intermedia* and *A. macrophthalma;* but I am at present unable to see any essential differences between the specimens so named and normal specimens of *A. cristata*. On the other hand, a very beautiful form, which has been collected by Dr. Grimm in great profusion from rather considerable depths in the middle and southern part of the Caspian Sea, and which was labelled *A. cristata*, var. *spinata*, would more likely seem to represent a distinct species.

4. Amathillina cristata, Grimm.

(Pl. V, Pl. VI, figs. 1—8).

Specific Characteristic. — Body rather stout and not very much compressed, with the back distinctly keeled throughout, the keel being, however, in its anterior part rather low, and scarcely elevated to any distinct projections in front of the 4th segment of mesosome, the succeeding projections successively increasing somewhat in size and being rather broad, triangular, that of last segment of metasome, however, differing from the others in being evenly rounded, not angulary produced. Cephalon with the rostral projection short and blunt, lateral lobes obtusely truncated. Anterior pairs of coxal plates somewhat deeper than the corresponding segments, 1st pair but slightly expanded distally, though considerably broader than the 2nd; 4th pair with the posterior expansion transversely truncated and forming below the emargination a nearly right angle. The last 2 pairs of epimeral plates of metasome but very slightly produced at the lateral corners. Eyes not very large, narrow reniform, with dark pigment. Superior antennae nearly equalling half the length of the body, joints of the peduncle successively diminishing in size, flagellum half as long again as the peduncle, accessory appendage about the length of the last peduncular joint and 5-articulate. Inferior antennae in female scarcely more than half as long as the superior. Gnathopoda in female comparatively small and about same length, propodos in the posterior ones considerably narrower than in the anterior, palm in both pairs somewhat oblique: those in male much stronger, with the propodos very large and somewhat claviform in shape, palm concave and defined below by an angular projecting lobe armed with 2 strong spines. Basal joint of antepenultimate and penultimate pairs of perciopoda of nearly same form, though somewhat differing in size, posterior edge in both pairs but slightly curved; that of last pair considerably broader in female than in male, posterior expansion forming below a rounded lobe reaching beyond the ischial joint.

The 2 anterior pairs of uropoda strongly spinous; last pair with the outer ramus somewhat longer than the basal part, its proximal joint having on either side a single fascicle of spines. Telson nearly semicircular in outline, cleft very narrow, each half armed with a lateral and an apical spine, the latter accompanied by a number of delicate bristles. Length of adult female 13 mm., of male 15 mm.

Remarks. — The present species, established by Dr. Grimm, may be regarded as the type of the genus *Amathillina*. It is chiefly distinguished by the number of the dorsal projections, and particularly by the peculiar, gibbous form of the last one, moreover by the shape of the basal joint of the last 2 pairs of pereiopoda.

Description of the female.

(Pl. V).

The length of fully adult ovigerous specimens amounts to about 13 mm.

The body (see figs. 1 and 2) is on the whole of a rather stout and compact form, being generally strongly curved. Its integuments are, however, not nearly so strongly incrusted as in the species of the 2 preceding genera, and do not exhibit any conspicuous sculpturing. In a dorsal view of the animal (fig. 2), the body appears much less compressed than in the species of the genus *Gmelina*, exhibiting a somewhat subfusiform shape, the greatest breadth (across the 4th segment of mesosome) equalling about $^{1}/_{5}$ of the length. The back is keeled throughout the whole mesosome and metasome; but the keel is in the anterior part rather low, becoming gradually more conspicuous posteriorly, where it is elevated in each segment to a lamellar, posteriorly pointing projection. The exact number of these dorsal projections is not easy to indicate, as they only little by little grow out from the segments. But in the 2 anterior segments of the mesosome there is never found any trace of such projections, and in the 3rd segment only in some specimens a slight attempt to a projection is observed. Not rarely even the dorsal projections are not at all distinctly formed in front of the 5th segment, that of the latter segment being in such cases rather small. In the last 2 segments of mesosome and those of metasome they, however, always appear well formed. The projection of the last segment of the metasome in all specimens distinguishes itself very markedly by its peculiar form, it being not, as in the preceding segments, triangular, but broadly rounded at the tip, giving that segment, in a lateral view of the animal, a somewhat gibbous appearance. The urosome is comparatively short and massive, without any dorsal keel or projections, but each of the segments carries dorsally a few simple hairs, and the 2 posterior ones have besides, on either side of the dorsal face, 2 small juxtaposed spinules, as in some species of the genus *Gammarus*.

The cephalon (see fig. 1, comp. also Pl. VI, fig. 2) scarcely exceeds in length the first 2 segments of mesosome combined, and is comparatively deep in proportion to its length, with the lateral faces quite smooth. The rostral projection is very short and blunt, though distinctly defined, and the lateral lobes are but little projecting and broadly truncated at the tip, being defined from the acutely projecting postantennal corners by a slight emargination encircling the basal joint of the inferior antennae.

The coxal plates are of moderate size, the 4 anterior pairs being, as usual, much larger than the 3 posterior, and somewhat deeper than the corresponding segments. The 1st pair (see fig. 11) are slightly expanded distally, and considerably broader in their outer part than the 2nd pair (see fig. 12), their terminal edge being broadly rounded and, as in the other pairs, only fringed with a few scattered hairs. The 2 succeeding pairs are somewhat deeper than the 1st and of oblong quadrangular form, the 3rd being somewhat broader than the 2nd. The 4th pair (see fig. 13) are much the largest, being rather expanded in their outer part and produced posteriorly to an obtusely truncated lobe, above which the posterior edge forms a distinct emargination, to receive the anterior part of the 5th pair. The latter (see fig. 14) are about twice as broad as they are deep, and, as usual, divided into 2 lobes, the anterior of which is but little larger than the posterior. The 2 posterior pairs (see figs. 15 and 16) successively decrease in size, and are also slightly bilobed.

The epimeral plates of the metasome are of moderate size, the 2 posterior pairs being, as usual, somewhat larger than the 1st pair, and nearly rectangular in form, with the lateral corners but little produced.

The eyes (see fig. 1), which are placed on the sides of the head, at a short distance from the anterior edges, are not very large and of a narrow reniform shape, with dark pigment.

The superior antennae (see fig. 1) nearly attain half the length of the body, and are rather slender, with only small scattered bristles at the edges. The peduncle is somewhat elongated, being about twice as long as the cephalon. The 1st joint is by far the largest, though scarcely as long as the other 2 combined, and the latter are not very different in length. The flagellum is about half as long again as the peduncle, and composed of numerous short articulations, their number varying from 20 to 25. The accessory appendage (see fig. 3) is well developed and about as long as the last peduncular joint, being composed of about 5 articulations.

The inferior antennæ (see fig. 1) are much shorter than the superior, scarcely exceeding half their length. They are constructed in the usual manner, exhibiting a large globular basal joint followed by two short and 2 elongated peduncular joints. The flagellum considerably exceeds half the length of the peduncle, and is composed of about 12 articulations.

The buccal area (see fig. 1) is rather projecting, being only partly obtected by the 1st pair of coxal plates. The several oral parts (figs. 4—10) composing it, are quite normal in their structure, and need not therefore to be described in detail.

The gnathopoda (figs. 11 and 12) are rather small and nearly of equal length, though the posterior ones (fig. 12) appear somewhat more slender than the anterior. Both pairs are rather richly supplied with bristles, partly arranged in dense fascicles, especially on the lower edge of the carpus and propodos. The latter appears in the anterior pair (fig. 11) somewhat broader and more expanded distally than in the posterior pair, where it (see fig. 12) exhibits a rather narrow oblong oval form. The palm in both pairs is somewhat oblique, being defined below by an obtuse angle carrying a pair of short spines.

The pereiopoda are of moderate length and rather strongly built, having their outer part edged with fascicles of stiff bristles intermingled with spines, especially at the end of the meral and carpal joints. In all of them the dactylus is very strong, terminating in a sharp curved point. The 2 anterior pairs are, as usual, of the same structure, though somewhat unequal in length, the 2nd pair (fig. 13) being a little shorter than the 1st.

Of the 3 posterior pairs the penultimate ones (fig. 15) are the longest, and have the basal joint oval in form, with the posterior edge but very slightly curved. In the antepenultimate pair (fig. 14) the basal joint is somewhat smaller, but otherwise of a much similar form, being in both pairs broadest in its proximal part and somewhat narrowed distally. The last pair (fig. 16) differ considerably from the others in the form of the basal joint, which is very broad, forming posteriorly a large, laminar expansion terminating below in a broadly rounded lobe which extends beyond the ischial joint. The edges of the expansion are minutely serrate, with small bristles springing of from the serrations, and having between them a very fine ciliation.

The branchial and inbubatory lamellæ (see fig. 12) exhibit a similar structure to that in the 2 preceding genera.

The uropoda successively decrease in size, the 1st pair (fig. 17) being rather large and about twice as long as the 2nd (fig. 18). In both pairs the basal part as also the rami are coarsely spinous, the latter being subequal and each tipped by a dense fascicle of unequal spines. The last pair (fig. 19) are very small, scarcely at all reaching beyond the others, and of a rather different structure. They consist each of a rather thick and massive basal part armed at the end with several spines, and of 2 very unequally developed rami. The inner ramus is extremely small and scale-like, whereas the outer is somewhat longer than the basal part and of a rather narrow, sublinear form, having a distinctly defined terminal joint setiferous at the tip. The proximal joint of this ramus carries on each side a single fascicle of spines intermingled with delicate bristles, and from its tip also issue several spines and fine bristles.

The telson (fig. 20) is rather broad in proportion to its length, and nearly semicircular in outline. It is divided by a deep and narrow cleft into two halves, each of which carries at the outer edge, near the base, a small spinule and at the tip another spinule accompanied by a few fine hairs.

The adult male (Pl. VI. fig. 1), as usual, attains a somewhat larger size than the female, its length amounting to nearly 15 mm.

The form of the body is not very much different form that in female, though perhaps a little more slender and compressed. The dorsal projections generally appear somewhat larger and more prominent, being more pronouncedly lamellar in character. In the specimen here figured there was a distinct attempt to such projections even in the 3rd and 4th segments of the mesosome, a case rather rarely met with, the projections being, as a rule, not distinctly developed in front of the 5th segment.

The antennae (see fig. 1) appear somewhat more elongated than in the female, especially the inferior ones, which however still are considerably shorter than the superior.

The gnathopoda (figs. 3 and 4) are very strongly developed and nearly equal in size, with the propodos rather large and of an oblong clavate form, being somewhat expanded distally, especially in the posterior ones (fig. 4). The palm is distinctly concave, and defined below by a projecting, nearly rectangular corner armed with 2 spines. Another rather strong spine occurs on the outher side of the palm below the middle, and is accompanied by a fascicle of slender bristles.

The pereiopoda appear a little more elongated than in the female, and the basal joint of the 3 posterior pairs is comparatively narrower. Especially is this the case with the last pair (fig. 6), where that joint appears much less expanded than in the female (comp. Pl. V. fig. 16) and thereby acquire a rather different form.

The last pair of uropoda (fig. 7) and the telson (fig. 8) do not differ much from those parts in the female.

Colour. — In some specimens received short time after they had been captured, a few light reddish markings were observed on the sides of the body, apparently being the remnant of a pigment; but whether this may have been something merely accidental, I cannot ascertain. In another bottle all the specimens exhibited along the edges of the dorsal projections a border of a very dark hue, as indicated in the figs. 1 and 2 on Pl. V.

Occurrence. — Of this characteristic form numerous specimens were collected by Mr. Warpachowsky in several localities of the North Caspian Sea. It has been noted from no less than 16 different Stations, distributed partly along the western coast, from the Bai Agrachansky up to the mouth of the Wolga, partly in the tract extending north of the peninsula Mangyschlack, and also in 2 Stations (31 and 32) lying about midways between the latter peninsula and the opposite western coast. In some of the Stations it would seem to have occurred in great profusion.

Dr. Grimm collected this form in the Bai of Baku, and besides in several Stations both of the southern and middle part of the Caspian Sea, up to the peninsula Mangyschlak; the depth varying from 2 to 35 fathoms. A small variety (perhaps a new species) was also collected by the same naturalist at Baku in comparatively shallow water, among the grass.

Out of the Caspian Sea this form has not yet been recorded.

5. Amathillina affinis, G. O. Sars, n. sp.

(Pl. VI. figs. 9—19).

Specific Characteristic. — Very like the preceding species, but of much inferior size. Anterior part of mesosome not keeled dorsally; the last 2 segments of mesosome and those of metasome each produced dorsally to a prominent, acutely triangular projection, that of last segment nearly of same form as the preceding ones. Cephalon and urosome almost as in *A. cristata.* Anterior pairs of coxal plates somewhat smaller than in the said species; otherwise of a similar shape. Eyes comparatively larger and distinctly reniform. Superior antennæ very slender and exceeding half the length of the body, 1st joint of the peduncle but little longer than the 2nd, accessory appendage shorter than the last peduncular joint, and only 3-articulate. Gnathopoda in female very small and of a similar structure to that in *A. cristata,* propodos of the posterior ones much narrower than that of the anterior and having the palm nearly transverse; those in male largely developed, with the propodos in both pairs oblong oval in form, scarcely widening distally. Basal joint of penultimate pair of pereiopoda very different in shape from that of the antepenultimate pair, being strongly expanded, with the posterior edge boldly curved below the middle; that of last pair having the posterior expansion produced below to an obtusely truncated lobe reaching almost to the middle of the meral joint. Uropoda nearly as in the preceding species. Telson without any spines, and having the terminal lobes obtusely pointed, each being tipped by 3 fine hairs. Length of adult female 6 mm., of male 8 mm.

Remarks. — This new species is very nearly allied to the preceding one, but unquestionably specifically distinct. Besides by its much inferior size, it differs in the anterior part of the back being quite smooth, without any trace of a keel, in the last dorsal projection not differing in shape from the preceding ones, and in the rather different form of the basal joint of the last 2 pairs of pereiopoda, finally, in the telson having no trace of any spines.

Description. — The length of adult, ovigerous female specimens is about 6 mm., and that of male specimens scarcely exceeds 8 mm. This form is consequently much inferior in sexe to the preceding species.

The general form of the body nearly agrees with that in *A. cristata*, and there is a quite similar difference between the 2 sexes as described in that species, the females being somewhat shorter and stouter than the males. On this cause I have regarded it sufficient for the recognition of the species to figure only one of the sexes, in this case the male (fig. 9). In both sexes the anterior part of the back is quite evenly rounded, without any trace of a keel. In the 5th segment of the mesosome there is found in some specimens a very slight approach to a keel, but in no specimen this keel is elevated in the form of a dorsal projection. In the 5 succeeding segments, on the other hand, the dorsal projections are very distinctly developed, being rather projecting and of an acutely triangular shape. The last of these projections does not differ much from the others, being, as the latter, acutely produced, not, as in the preceding species, rounded. The segments of the urosome are, as in that species, without any dorsal keel or projections, but provided with a similar supply of fine hairs and small subdorsal spinules.

The cephalon (fig. 10) does not differ much in its form from that in *A. cristata*.

The coxal plates are comparatively somewhat less deep than in the preceding species and also narrower, otherwise of a much similar appearance.

This also applies to the epimeral plates of the metasome.

The eyes (see fig. 10) are comparatively larger than in *A. cristata*, and of a pronounced reniform shape, their anterior edge being distinctly insinuated in the middle.

The superior antennæ (see fig. 9) are very slender, and considerably exceed in length half the body. The 1st joint of the peduncle does not much exceed in length the 2nd, and the 3rd joint is considerably both shorter and narrower than the latter. The flagellum is about half again as long as the peduncle, and composed in the female of about 15 articulations, in the male of nearabout the double number. The accessory appendage in both sexes is much smaller than in *A. cristata*, and is only composed of 3 articulations.

The inferior antennæ are much shorter than the superior, especially in the female, and of a similar structure as in *A. cristata*.

The gnathopoda in the female (figs. 11 and 12) are rather small and nearly of equal length, though the posterior ones appear somewhat feebler in structure. The propodos in the latter (fig. 12) is much narrower than in the anterior, and has the palm nearly transverse. In the male these limbs (figs. 18 and 19) are very strongly developed and of a similar structure to that in the male of the preceding species, though differing in the propodos being more regularly oval in form, that of the anterior pair (fig. 1) being rather tumid in the middle.

Of the pereiopoda, the last 2 pairs differ very markedly from those of the preceding species in the shape of the basal joint. In the penultimate pair (fig. 14) this joint is very unlike that of the antipenultimate pair (fig. 13), forming a large and broad expansion posteriorly, whereby it acquires a somewhat heart-shaped form, the posterior edge being boldly curved below the middle. In the last pair (fig. 15) it expands obliquely to a greatly projecting lobe, obtusely truncated at the tip and extending almost to the middle of the meral joint. In the male these joints are somewhat less expanded than in the female, being however much broader than in the male of *A. cristata*.

The uropoda are nearly of same structure as in that species, except that the 2 anterior pairs are armed with a less number of spines, and that the outer ramus of the last pair (fig. 16) is somewhat more elongated.

The telson (fig. 17) has the terminal lobes obtusely pointed and each only tipped by 3 fine hairs, no spine being found neither on the tip nor on the outer edge.

Occurrence. — This species also has been collected by Mr. Warpachowsky in several localities of the North Caspian Sea, it being noted from no less than 11 different Stations, but in none of them it occurred in any abundance. Of these Stations one (St. 2) is located off the Tschistyi-Bank, another (St. 12) in the inner part of the Bai Agrachansky, 4 other (St. 16, 17, 28, 29) in the tract north of the peninsula Mangyschlak, an 8th (St. 32) about midway between that peninsula and the opposite western coast, another (St. 49) between the islands Morskay and Kulaly, and the last 3 (St. 54, 55, 56) at some distance north and west of the last-named island.

In the collection of Dr. Grimm this species is only represented by a few specimens collected in the Bai of Baku, from a depth of 2—3 fathoms.

The species is, as yet known, restricted in its occurrence to the Caspian Sea.

Gen. 4. **Gammarus**, Fabr.

Remarks. — Of all the Amphipodous genera represented in the Caspian Sea, this comprises the greatest number of species. In the collection of Mr. Warpachowsky I have distinguished no less than 11 different species, and in the collection of Dr. Grimm several additional species are represented. Whereas the hitherto known species of *Gammarus*, in the restriction of the genus now generally adopted, exhibit a very uniform appearance, the Caspian species partly diverge rather markedly in their character from the type, both as regards the outward appearance and the structure of the several appendages. Thus the *Gammarus caspius* Pallas, to be described below, is highly distinguished by the segments of metasome being produced dorsally to similar acuminate projections to those occurring in the genus *Amathillina*, and whereas in the earlier known species of *Gammarus*, the superior antennae are invariably very slender and considerably longer than the inferior, in several of the Caspian species they are rather much reduced in length, so as not at all exceeding the inferior ones in size. Moreover the last pair of uropoda sometimes are unusually short, and in all the Caspian species as yet examined their inner ramus is very small and scale-like. The most normally looking species is that described below as *Gammarus haemobaphes* Eichwald.

6. **Gammarus caspius**, Pallas.

(Pl. VII).

Gammarus caspius Pall., Eichwald: «Fauna caspio-caucasia nonnullis observationibus novis illustr.». Nouv. Mém. de la Soc. Imp. des Naturalistes de Moscou, T. VII, 1842, p. 230.

Syn.: *Gammarus semicarinatus.* Sp. Bate.
» *Gammarus Dybowskyi,* Grimm MS.

Specific Characteristic. — Body moderately slender, with the segments of mesosome generally smooth, though in some specimens the last one is slightly keeled above and produced at the posterior edge to a small dentiform projection, those of metasome provided with well-marked posteriorly pointing dorsal projections. The 2 anterior segments of urosome having each a much elevated tubercle, transversely truncated at the tip and armed with 4 strong apical spines arranged in pairs; last segment with a single small spinule on each side of the dorsal face. Cephalon with the rostral projection extremely small, nearly obsolete, lateral lobes rather broad and obtusely truncated at the tip. Anterior pairs of coxal plates but little deeper than the corresponding segments and rapidly increasing in size to the 4th, which are much expanded in their outer part, with a very distinct emargination posteriorly. The last 2 pairs of epimeral plates of metasome rather large and acutely produced at the lateral corners. Eyes well developed and of an oblong form, slightly instricted in the middle. Superior antennæ very slender and much longer than the inferior, joints of the peduncle rapidly diminishing in size, flagellum nearly twice as long as the peduncle, accessory appendage well developed and 5-articulate. Gnathopoda in both sexes rather unequal in size, the posterior ones being much the larger; those in male being, as usual, more powerful than in female, with the propodos rather large, especially in the posterior ones, palm in both pairs somewhat oblique and nearly straight. Pereiopoda moderately slender and edged in their outer part with spines and delicate bristles, antepenultimate pair much shorter than the last 2 pairs, which are nearly equal in length, basal joint of last pair not much expanded and oblong quadrangular in form, with the posterior edge distinctly serrate. Last pair of uropoda reaching considerably beyond the other, inner ramus small, squamiform, outer ramus rather elongated and edged with long ciliated setæ and a few fascicles of spines. Telson of moderate size and cleft to the base, each half armed at the tip with 2 small spines and a few delicate bristles. Length of adult female 13 mm., of male 16 mm.

Remarks. — The diagnosis given by Eichwald in the above-cited work does not leave any doubt, that the above-characterised form is that originally recorded by Pallas as *Gammarus caspius*. Under the latter name Sp. Bate, in his Catalogue of Amphipoda in the British Museum, describes a very different form, whereas I am much inclined to believe that the form recorded by him in the same work (without any locality) as *G. semicarinatus* is that here treated of. In Dr. Grimm's collection this species is labelled *G. Dybowskyi* n. sp. From all other known species this is at once recognized by the strong dorsal projections of the metasome. In spite of this anomalous feature, it is a true *Gammarus*, as shown by the structure both of the oral parts and the other appendages.

Description of the female.

The length of adult ovigerous specimens amounts to about 13 mm.

The body (see fig. 1) is of moderately slender form and somewhat compressed, with the metasome and urosome well developed and combined about equalling the length of the mesosome. The segments of the latter division are in most of the specimens quite smooth, with the back evenly rounded. In larger specimens there is however (as indicated in the figures here given) not rarely found in the last segment a slight dorsal keel, which at the posterior edge is produced to a small dentiform projection. The segments of metasome in all specimens are distinctly keeled, the keel being elevated to rather large and compressed, posteriorly pointing dorsal projections terminating in a very acute point. The last of these projections is generally the largest and of same form as the 2 preceding ones. The 2 anterior segments of the urosome are each provided dorsally with a rather conspicuous, almost cylindrical tubercle, transversely truncated at the tip, and carrying 4 strong apical spines arranged in pairs and accompanied by a few delicate bristles (see fig. 15). The anterior tubercle projects nearly at a right angle to the longitudinal axis, whereas the posterior one is slightly recurved, both being otherwise of the very same appearance. The last segment of the urosome has on each side of the dorsal face a single small spinule.

The cephalon (fig. 2) is fully as long as the first 2 segments of mesosome combined, and has the rostral projection extremely small, nearly obsolete. The lateral lobes are somewhat projecting and rather broad, being obtusely truncated at the tip and defined from the acutely produced post-antennal corners by a rather deep emargination encircling the globular basal joint of the inferior antennæ.

The 4 anterior pairs of coxal plates are but little deeper than the corresponding segments, and rapidly increase in size posteriorly, the 1st pair (see fig. 4) being rather small and scarcely at all expanded distally, whereas the 4th pair (see fig. 6) are very broad, with the outer part much expanded and forming below the rather deep posterior emargination a distinct, almost right angle.

The 3 posterior pairs of coxal plates are comparatively small and of the usual shape.

The epimeral plates of the metasome are rather large, especially the 2 posterior pairs, which both are produced at the lateral corners to an acute point.

The eyes (see fig. 2) are of moderate size and narrow oblong in form, with a slight instriction in the middle, thus exhibiting a shape somewhat similar to that in the northern species, *G. campylops* Leach. The pigment in most of the specimens is dark, but Dr. Grimm has stated a case of the eyes being nearly devoid of pigment.

The superior antennæ (see fig. 1) about equal half the length of the body, and are very slender, with only scattered short hairs at the edges. The joints of the peduncle rapidly diminish in size, the 1st being much the largest and about equalling in length the other 2 combined. The last peduncular joint is considerably shorter and also narrower than the 2nd. The flagellum does not fully attain twice the length of the peduncle, and is composed of numerous short articulations. The accessory appendage (fig. 3) is well developed, somewhat longer than the last peduncular joint, and composed of 5 articulations.

The inferior antennæ are much shorter than the superior, but little exceeding half their length, and have the penultimate joint of the peduncle the largest. The flagellum somewhat exceeds half the length of the peduncle, and is composed of about 9 articulations.

The oral parts do not differ in any way from those in the other species of *Gammarus*.

The gnathopoda (figs. 4 and 5) are moderately strong and rather unequal in size, the posterior ones (fig. 5) being much the larger. In both pairs the carpus is rather short and expanded distally, forming below a rounded, setiferous lobe. The propodos is in the posterior ones considerably larger than in the anterior, but of a similar form in both pairs, being oval quadrangular in shape, with the palm somewhat oblique, and defined below by an obtuse angle carrying a strong spine.

The pereiopoda are of moderate length and have their outer part edged with fascicles of short spines and delicate bristles. The 2 anterior pairs (see fig. 6) are rather slender and somewhat unequal in length, the 1st pair being the longer. The antepenultimate pair (fig. 7) are considerably shorter than the 2 succeeding pairs, and have the basal joint of an irregular oval form, with the infero-posteal corner slightly produced. The last 2 pairs are about equal in length, but differ in the shape of the basal joint, which in the last pair (fig. 8) is somewhat larger than in the penultimate pair, though not very much expanded, exhibiting an oblong quadrangular form, and having the posterior edge, as in the 2 preceding pairs, distinctly serrate.

The 2 anterior pairs of uropoda (figs. 9 and 10) are of the usual structure, the rami being linear in form and nearly equal-sized. They are edged with a number of coarse spinules and have each at the tip a fascicle of somewhat unequal spines.

The last pair of uropoda (fig. 11) considerably project beyond the others, and have the basal part armed at the end below with 4 strong juxtaposed spines. The inner ramus is very small and scale-like, carrying a single small spine at the tip and another still smaller on the inner edge. The outer ramus is well developed and nearly 3 times as long as the basal part. It is comparatively narrow, slightly tapering distally, and is provided at the tip with a very small terminal joint. The ramus is round about edged with long ciliated setæ, and besides exhibits a few fascicles of short spines, 2 of which issue from the tip, on either side of the terminal joint.

The telson (fig. 12) is not very large, and scarcely extends beyond the basal part of the last pair of uropoda. It is divided by a deep cleft into two halves, each slightly narrowed distally and carrying at the somewhat obliquely truncated tip 2 small spines and a few fine hairs.

The adult male (fig. 13), as usual, grows to a somewhat larger size than the female, the largest specimens measuring about 16 mm. in length.

In its general form the body does not differ much from that in the female, being only a little more slender and compressed, and having the coxal plates comparatively smaller.

The antennæ appear somewhat less unequal, the inferior ones being comparatively more fully developed than in the female and also more densely setiferous. The accessory appendage of the superior ones (see fig. 14) is a little more elongated than in the female, though exhibiting the same number of articulations.

The gnathopoda (figs. 16 and 17) are much stronger than in the female and, as in the latter, rather unequal in size, the posterior ones (fig. 17) being considerably more powerful than the anterior. In both pairs the propodos exhibits a similar oval quadrangular form to that in the female, but is much larger, especially that of the posterior pair. The palm is nearly straight and somewhat oblique, being defined below by an obtuse angle carrying 2 strong spines, between which the dactylus impinges, when closed; besides the palm has on the outer side, about in the middle, a strong spine, not occurring in the female.

The pereiopoda (see fig. 13) appear somewhat more slender than in the female, and the basal joint of the 3 posterior pairs is also comparatively narrower.

The last pair of uropoda (fig. 18) are a little more elongated than in the female, nearly equalling in length the urosome, but otherwise are of a much similar structure.

Colour. — In none of the specimens examined any colouring marks could be detected, the whole body exhibiting a uniform whitish hue.

Occurrence. — This form was collected rather abundantly by Mr. Warpachowsky in the North Caspian Sea, and has been noted from no less than 16 different Stations. Of these one (St. 2) is located off the Tschistyi-Bank, another (St. 12) in the inner part of the Bai Agrachansky, a third (St. 31) about midway between the peninsula Mangyschlak and the opposite western coast, the others in the tract north of the said peninsula, 2 of them (St. 53 and 54) lying at some distance north of the islands Kulaly and Morskay. In some of the Stations, especially in St. 52 (off the island Swjatoj), it would seem to have occurred in great profusion.

Dr. Grimm collected this species in the Bai of Baku, from a depth of 4 feet down to 6 fathoms, furthermore in the Bai Balchansky, 7—12 fms., in the Bai Murrawjew, 10—20 fms., and on the west coast of Sara, among Zostera. A single specimen in the collection was, according to the label, taken by Kessler at Astrachan from *Astacus leptodactylus*. The specimens in the collection of Dr. Grimm are on the whole of much smaller size than those collected by Mr. Warpachowsky in the North Caspian Sea.

According to Eichwald, this form was collected by Pallas in the mouth of «Rhymnus» together with *G. pulex* (= *G. hæmobaphes*).

Out of the Caspian Sea it has not yet been recorded.

7. Gammarus hæmobaphes, Eichwald.

(Pl. VIII).

Gammarus hæmobaphes, Eichwald l. c. p. 230, Pl. XXXVII, fig. 7.
Syn.: *Gammarus pulex*, Pallas (not Fabr.).

Specific Characteristic. — Body resembling in form that in the more typical Gammari (e. g. *G. locusta*), being rather slender and compressed, with the mesosome and metasome perfectly smooth throughout. The 2 anterior segments of urosome each having a small, conical dorsal tubercle tipped by 2 minute juxtaposed spines; 1st segment besides provided, on each side of the dorsal face, with a single small spinule, and last segment with 2 such spinules. Cephalon with the lateral lobes rather broad and somewhat obliquely truncated at the tip, the inferior corner being more prominent than the superior. Coxal plates of moderate size, 4th pair rather broad in their outer part, and angularly produced below the posterior emargination. Last pair of epimeral plates of metasome but very little produced at the lateral corners. Eyes well developed, reniform, pigment dark. Superior antennæ rather slender and longer than the inferior, with the accessory appendage rather fully developed, and composed of 7—9 articulations. Gnathopoda in both sexes, very unequal in size, the posterior ones being much stronger than the anterior, and in male very powerful, with the propodos exceedingly large and swollen. The 2 anterior pairs of pereiopoda normal, the 3 posterior pairs rather stout, with their outer part edged with fascicles of strong spines and scattered bristles, basal joint of antepenultimate pair having the infero-posteal corner slightly produced, that of last pair much larger than in the preceding pairs, and subquadrangular in form, being broader in female than in male and in both sexes produced at the infero-posteal corner to a short, narrowly rounded lobe, posterior edge distinctly serrate. Last pair of uropoda reaching considerably beyond the others, and having the inner ramus small, scale-like, the outer elongated and densely fringed with ciliated setæ. Telson comparatively small, each half having at the tip one or two small spinules. Length of adult female 15 mm., of male 16 mm.

Remarks. — In all essential points the description and figures given by Eichwald of his *G. hæmobaphes* would seem to accord with the species above characterised, though they certainly are not detailed enough to give full evidence of the identity of both. The description of Eichwald, it is true, was made out from specimens collected in the Black Sea, but he believe that the same species also occurs in the Caspian Sea and that the form recorded by Pallas as *G. pulex* is most probably the same. As indeed several species both of *Mysidæ, Cumacea* and *Amphipoda* have been stated to be common to the two Seas, I cannot see any reason, why not the same could be the case with the present species. In every case there is but little chance of believing that the name proposed by Eichwald should be restored by other authors, and it may thus be properly applied to the form in question. The species may be best distinguished from the earlier known forms by the armature of the urosome and the rudimentary condition of the inner ramus of the last pair of uropoda, as also by the structure of the gnathopoda in the two sexes.

A form very nearly allied to the one here treated of has been collected by Dr. Grimm in great profusion in the southern and middle part of the Caspian Sea, partly from very considerable depths. This form, which has been named by that naturalist *Gammarus robustus*[1]), may perhaps turn out to be only a variety of the present species, though it differs markedly by its larger size, the more slender form of the several appendage, and by the shape of the dorsal tubercles of the urosome, which are developed nearly in a similar manner to that in *G. caspius.*

---

[1]) This name has been preoccupied in the year 1875 by Prof. S. Smith for a North-American species.

Description of the female.

The largest female specimens in the collection of Mr. Warpachowsky reach a length of 15 mm., but there are also fully adult ovigerous specimens of much inferior size.

In its general appearance (see fig. 1) the animal looks very like the well known typical species, *G. marinus, locusta* and *pulex*. As in the latter, the body appears rather slender and compressed, with the mesosome and metasome quite smooth throughout and the back evenly rounded, without any trace of keel or projections. The urosome (see also fig. 3) is of moderate size, and has the 2 anterior segments each elevated dorsally to a small conical tubercle carrying at the tip 2 minute, juxtaposed spinules accompanied by a pair of fine hairs. Besides the 1st segment has on each side of the dorsal face a single spinule, and 2 such spinules occur on the same place in the last segment.

The cephalon (fig. 2) about equals in length the first 2 segments of mesosome combined, and appears almost transversely truncated at the tip, the rostral projection being extremely small. The lateral lobes are rather broad and somewhat obliquely truncated, with the inferior corner the more prominent. They are defined form the postantennal corners by a very deep, nearly angular emargination encircling the greatly swollen basal joint of the inferior antennæ.

The 4 anterior pairs of coxal plates are of moderate size, being somewhat deeper than the corresponding segments, and successively increase in size posteriorly. The 3 anterior pairs are nearly quadrangular in shape, whereas the 4th pair exhibit a rather irregular form, having their outer part considerably expanded and angularly produced below the posterior emargination.

The 3 posterior pairs of coxal plates are comparatively small and of the usual shape.

The epimeral plates of the metasome are well developed, the 2 posterior pairs being, as usual, larger than the anterior pair and both but very slightly produced at the lateral corners.

The eyes (see fig. 2) are of moderate size and of a pronouncedly reniform shape, with well developed visual elements and dark pigment.

The superior antennæ (see fig. 1) nearly attain half the length of the body, and are rather slender and but very sparingly setiferous. The joints of the peduncle successively diminish in size, the 1st being much the largest and equalling in length the other 2 combined. The flagellum is nearly twice as long as the peduncle, and composed of numerous short articulations. The accessory appendage (fig. 5) is rather fully developed, equalling half the length of the peduncle, and is composed of about 7 articulations.

The inferior antennæ, as in most of the typical Gammari, are shorter than the superior and somewhat more densely setiferous. The last 2 joints of the peduncle are nearly equal-sized and combined somewhat longer than the flagellum, which is composed of about 8 articulations.

The gnathopoda (figs. 6 and 7) are rather unequal in size, the posterior ones (fig. 7) being much stronger than the anterior. In structure they agree rather closely with those in the female of the preceding species, the carpus being in both pairs comparatively short and expanded distally, with a rounded setiferous lobe below. The propodos in both pairs considerably exceeds in length the 3 preceding joints combined, and in the posterior pair is much larger and more tumid than in the anterior. The palm is somewhat oblique and defined below by an obtuse angle carrying a strong spine followed by a few much shorter ones. The hind margin of the propodos in both pairs is provided with numerous small tufts of bristles.

Of the pereiopoda, the 2 anterior pairs (fig. 8) exhibit the usual slender form. The 3 posterior pairs are, on the other hand, rather stout and have their outer part edged with fascicles of strong spines and scattered bristles. As usual, the antepenultimate pair (fig. 9) are considerably shorter than the 2 succeeding ones, and have the basal joint of a somewhat irregular quadrangular form, with the infero-posteal corner nearly rectangular. In the penultimate pair (fig. 10) the basal joint is somewhat larger and more expanded in its proximal part, the posterior edge being boldly curved above and not at all produced at the infero-posteal corner. The last pair (fig. 11) about equal in length the penultimate pair, and have the basal joint much larger than in any of the preceding pairs and of a rounded quadrangular shape, forming posteriorly a broad laminar expansion, which terminates below in a short, narrowly rounded lobe. The posterior edge of the expansion is slightly curved and, as in the 2 preceding pairs, exhibits a number of distinct serrations, each carrying a small hair.

The 2 anterior pairs of uropoda (figs. 12 and 19) are normal in structure, though less coarsely spinous than in the preceding species, their inner ramus having only a single lateral spine and the outer no lateral spines at all.

The last pair of uropoda (fig. 13) considerably project beyond the others, and on the whole agree in their structure with those in the preceding species; the inner ramus being very small and scale-like, whereas the outer is rather elongated and densely edged with long ciliated setæ, and having besides a few fascicles of short spines. The terminal joint of the ramus is very small and nearly hidden between the spines issuing from the tip of the proximal joint.

The telson (fig. 14) is comparatively small, being scarcely as long as it is broad at the base. It is, as usual, divided by a deep cleft into 2 halves, each of which is somewhat narrowed in its outer part and armed with a single small apical spine accompanied by a pair of simple hairs.

The adult male (fig. 15) is generally somewhat larger than the female, reaching a length of about 16 mm. The body does not differ much in its general form from that in the female, except in being somewhat more compressed, and having the coxal plates less deep.

The antennæ are, as usual, somewhat more fully developed than in the female, and especially the inferior ones more strongly built and generally also more densely setiferous. The accessory appendage of the superior ones (fig. 16) appears more elongated and is composed of a greater number of articulations amounting to 9 in all.

The gnathopoda are still more unequally developed than in the female, the anterior ones (fig. 17) chiefly differing from those in the latter by the propodos being somewhat more elongated. The posterior gnathopoda (fig. 18), on the other hand, are of quite an unusual size, the propodos being exceedingly large, nearly occupying the half length of the leg. It is of a somewhat obpyriform shape, being not fully twice as long as it is broad, and, as in the female, has the palm rather oblique and quite straight, without any lateral spine in the middle. The hind margin is in some specimens very densely setous, and the dactylus is strong and curved.

The pereiopoda are perhaps a little more slender than in the female, and the basal joint of the 3 posterior pairs somewhat narrower.

The last pair of uropoda (see fig. 15) are, as usual, more fully developed than in the female, attaining about the length of the urosome, and have the marginal setæ of the outer ramus longer and more coarsely ciliated.

The telson (fig. 20) is of the very same shape as in the female; but generally 2, instead of a single spine, are found on the tip of each of the terminal lobes.

Colour. — According to Eichwald, the body, in the living state of the animal, exhibits a brownish green colour, the posterior edges of the segments being on each side tinged with pink.

Occurrence. — This form has been collected by Mr. Warpachowsky in 7 different Stations of the North Caspian Sea, but in none of the Stations it would seem to have occurred in any abundance. Of the Stations 2 (St. 16 and 17) are located off the island Swjatoj, a third (St. 24) between the islands Kulaly and Morskoy, 2 others (St. 31 and 32) about midway between the peninsula Mangyschlak and the opposite western coast, another (St. 40) north of the promontory Kossa Brjanskaja, the last, finally (St. 63), in the eastern part of the North Caspian Sea.

Besides, some specimens preserved in the Museum of St. Petersburgh from older time, and collected by Goebel and v. Baer partly at Baku, partly at the island Sara, would seem to be referable to this species.

Typical specimens of this form have been collected by Dr. Grimm at Baku in comparatively shallow water, as also in the middle part of the Caspian Sea, from the shores down to 40 fathoms.

Distribution. — The Black Sea (Eichwald).

EXPLANATION OF THE PLATES.

Pl. I.

Boeckia spinosa, Grimm.

Fig. 1. Adult female, viewed from left side.
» 2. Same, dorsal view.
» 3. Left superior antenna.
» 4. Left inferior antenna.
» 5. Urosome with its appendages, viewed from left side.
» 6. Telson viewed from above.
» 7. Last uropod.
» 8. Anterior lip.
» 9. Posterior lip.
Fig. 10. Left mandible with palp.
» 11. Right mandible, without the palp.
» 12. First pair of maxillæ.
» 13. Second maxilla.
» 14. Maxillipeds, without the right palp.
» 15. Left anterior gnathopod, with the corresponding coxal plate.
» 16. Left posterior gnathopod, with the corresponding coxal plate, branchial and incubatory lamella.

Pl. II.

Boeckia spinosa, Grimm.
(continued).

Fig. 1. Adult male, dorsal view.
» 2. Left anterior gnathopod, with the corresponding coxal plate.
» 3. Left posterior gnathopod, with the corresponding coxal plate and branchial lamella.
» 4. Second perciopod with the corresponding coxal plate.
Fig. 5. Antepenultimate perciopod, with coxal plate and branchial lamella.
» 6. Penultimate perciopod.
» 7. Last perciopod.
» 8. Second uropod.
» 9. Last uropod.
» 10. A very young specimen, viewed from left side.

Pl. III.

Gmelina costata, Grimm.

Fig. 1. Adult female, viewed from left side.
» 2. Cephalon with the base of the left inferior antenna, lateral view.
» 3. Part of the right superior antenna, showing the accessory appendage and the base of the flagellum.
» 4. Anterior lip.
» 5. Posterior lip.
» 6. Right mandible, without the palp.
» 7. Left mandible with palp.
» 8. First maxilla.
» 9. Second maxilla.
» 10. Maxillipeds, without the right palp.
Fig. 11. Left anterior gnathopod, with the corresponding coxal plate.
» 12. Left posterior gnathopod.
» 13. Last perciopod.
» 14. Second uropod.
» 15. Last uropod.
» 16. Telson, from above.
» 17. Adult male, viewed from right side.
» 18. Same, dorsal view.
» 19. Right anterior gnathopod.
» 20. Right posterior gnathopod.
» 21. First uropod.
» 22. Last uropod.

Pl. IV.

Gmelina Kusnezowi, (Sowinsky).

Fig. 1. Adult female, viewed from left side.
» 2. Cephalon with the base of the left inferior antenna, lateral view.
» 3. Part of the right superior antenna, showing the accessory appendage and the base of the flagellum.
Fig. 4. Left anterior gnathopod, with part of the corresponding coxal plate.
» 5. Left posterior gnathopod.
» 6. First perciopod.
» 7. Antepenultimate perciopod, with coxal plate.

Fig. 8. Last pereiopod.
" 9. Second uropod.
" 10. Last uropod.
" 11. Telson, from above.
" 12. Adult male, viewed from right side.

Fig. 13. Same, dorsal view.
" 14. Right anterior gnathopod.
" 15. Right posterior gnathopod.
" 16. Last uropod.
" 17. Telson, from above.

Pl. V.

Amathillina cristata, Grimm.

Fig. 1. Adult female, viewed from left side.
" 2. Same, dorsal view.
" 3. Part of the left superior antenna, showing the accessory appendage and the base of the flagellum.
" 4. Anterior lip.
" 5. Posterior lip.
" 6. Left mandible, without the palp.
" 7. Right mandible with palp.
" 8. First maxilla.
" 8a (not numbered in the plate). Palp of the right maxilla of same pair.
" 9. Second maxilla.
" 10. Maxillipeds, without the right palp.

Fig. 11. Left anterior gnathopod, with the corresponding coxal plate.
" 12. Left posterior gnathopod, with coxal plate, branchial and incubatory lamellæ.
" 13. Second pereiopod with coxal plate.
" 14. Antepenultimate pereiopod.
" 15. Penultimate pereiopod.
" 16. Last pereiopod.
" 17. First uropod.
" 18. Second uropod.
" 19. Last uropod.
" 20. Telson.

Pl. VI.

Amathillina cristata, Grimm,

(continued).

Fig. 1. Adult male, viewed from left side.
" 2. Cephalon with the base of the left inferior antenna, lateral view.
" 3. Left anterior gnathopod with coxal plate.

Fig. 4. Left posterior gnathopod.
" 5. Base of penultimate pereiopod.
" 6. Last pereiopod.
" 7. Last uropod.
" 8. Telson.

Amathillina affinis, G. O. Sars.

Fig. 9. Adult male, viewed from right side.
" 10. Cephalon of a female specimen, lateral view.
" 11. Right anterior gnathopod of female, with the corresponding coxal plate.
" 12. Right posterior gnathopod of same, with coxal plate, branchial and incubatory lamellæ.

Fig. 13. Antepenultimate pereiopod.
" 14. Penultimate pereiopod.
" 15. Last pereiopod.
" 16. Last uropod.
" 17. Telson.
" 18. Right anterior gnathopod of a male specimen.
" 19. Right posterior gnathopod of same.

Pl. VII.

Gammarus caspius, Pallas.

Fig. 1. Adult female, viewed from left side.
" 2. Cephalon with the base of the right inferior antenna, lateral view.
" 3. Accessory appendage of a superior antenna.
" 4. Left anterior gnathopod, with coxal plate.
" 5. Left posterior gnathopod, with coxal plate, branchial and incubatory lamellæ.
" 6. Second pereiopod with coxal plate.
" 7. Antepenultimate pereiopod.
" 8. Last pereiopod.
" 9. First uropod.

Fig. 10. Second uropod.
" 11. Last uropod.
" 12. Telson.
" 13. Adult male, viewed from right side.
" 14. Part of the left superior antenna, showing the last peduncular joint, the accessory appendage, and the base of the flagellum.
" 15. Part of the 2 anterior segments of urosome, showing the dorsal tubercles, lateral view.
" 16. Right anterior gnathopod.
" 17. Right posterior gnathopod, without the proximal part of the basal joint.

Pl. VIII.

Gammarus hæmobaphes, Eichwald.

Fig. 1. Adult female, viewed from left side.
» 2. Cephalon with the base of left inferior antenna, lateral view.
» 3. Urosome with telson, but without the uropoda, lateral view.
» 4. Lateral corner of last epimeral plate of metasome.
» 5. Accessory appendage of a superior antenna.
» 6. Left anterior gnathopod, with coxal plate.
» 7. Left posterior gnathopod, with coxal plate, branchial and incubatory lamellæ.
» 8. First pereiopod.
Fig. 9. Antepenultimate pereiopod.
» 10. Base of penultimate pereiopod.
» 11. Last pereiopod.
» 12. Second uropod.
» 13. Last uropod.
» 14. Telson.
» 15. Adult male, viewed from right side.
» 16. Accessory appendage of a superior antenna.
» 17. Right anterior gnathopod, with coxal plate.
» 18. Right posterior gnathopod.
» 19. First uropod.
» 20. Telson.

G.O.Sars Crustacea caspia.
Amphipoda Pl. I.

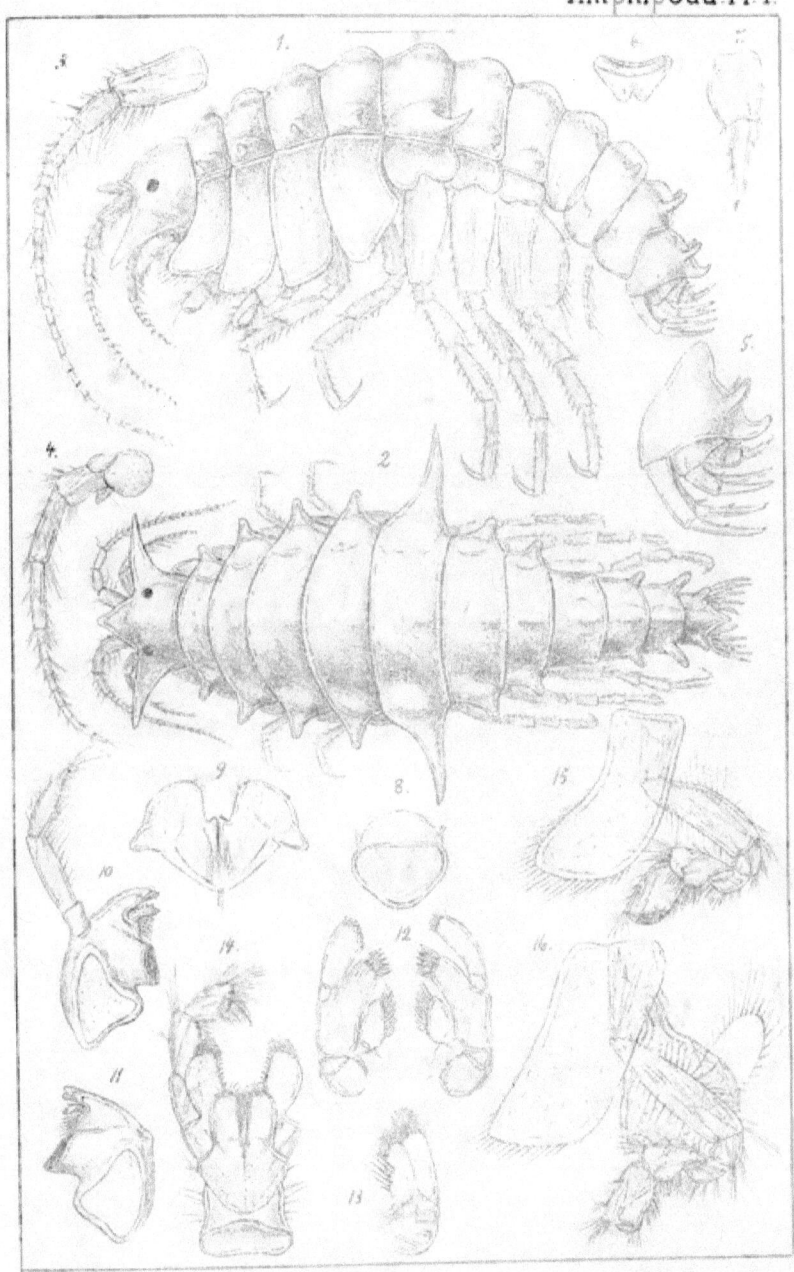

G.O.Sars autogr.

Boeckia spinosa, Grimm.

G.O.Sars Crustacea caspia.
Amphipoda. Pl. II.

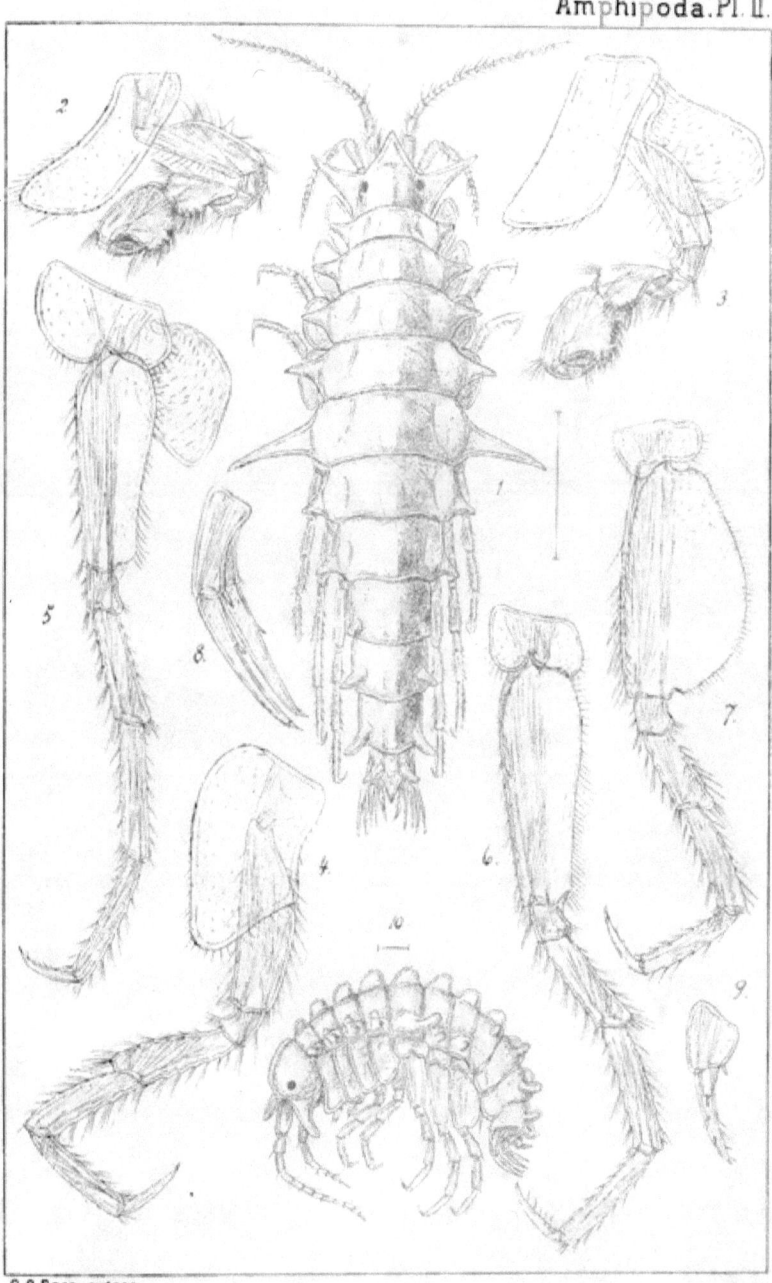

G.O.Sars autogr.

Boeckia spinosa, Grimm. (contin.)

G.O.Sars Crustacea caspia.
Amphipoda. Pl III.

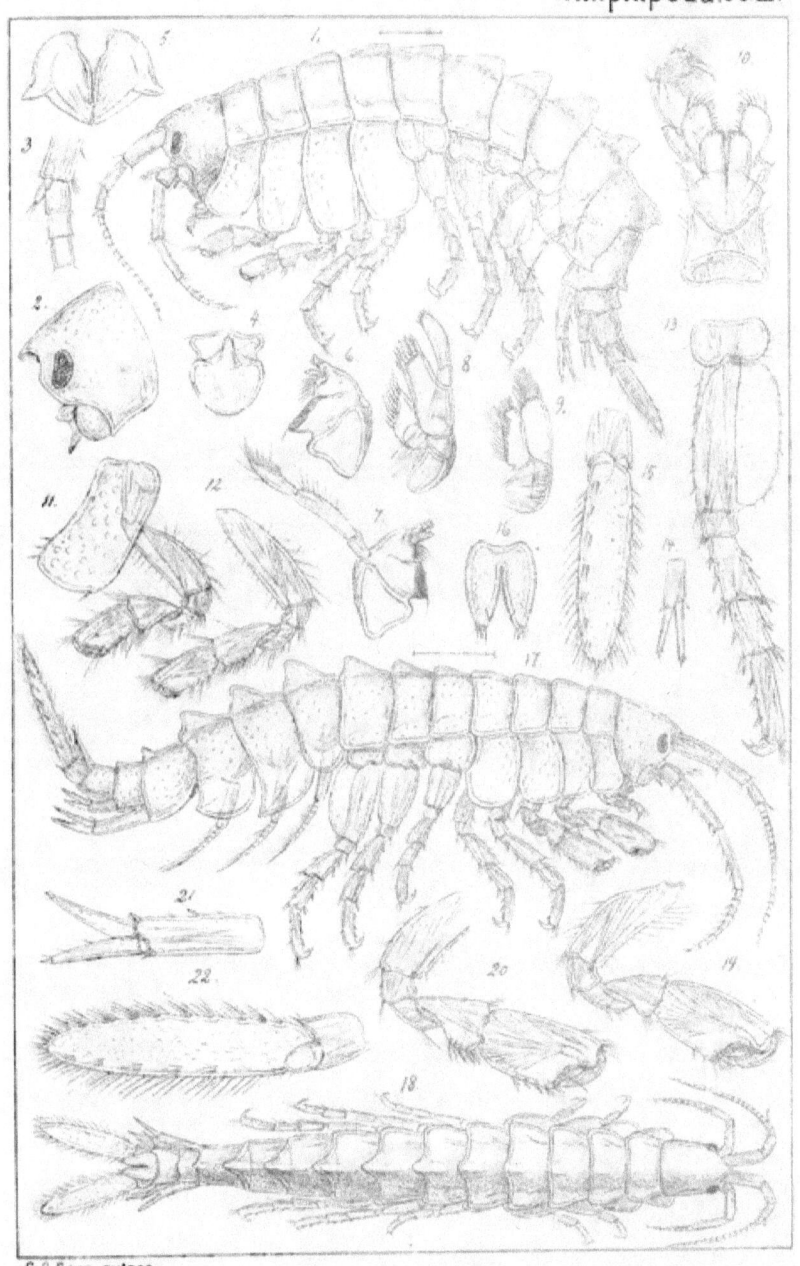

Gmelina costata, Grimm.

G.O.Sars Crustacea caspia. Amphipoda. Pl. IV.

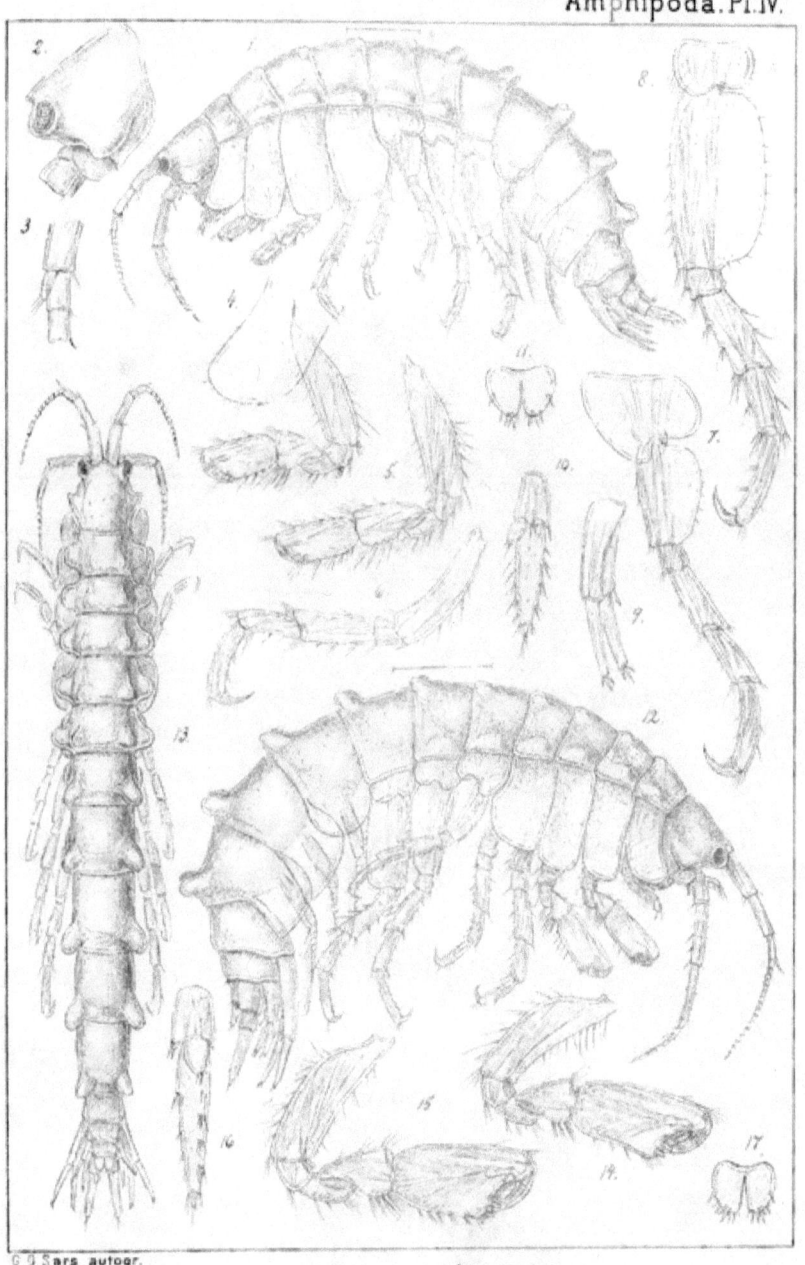

Gmelina Kusnezowi, (Sowinsky)

G.O.Sars Crustacea caspia.
Amphipoda. Pl. V.

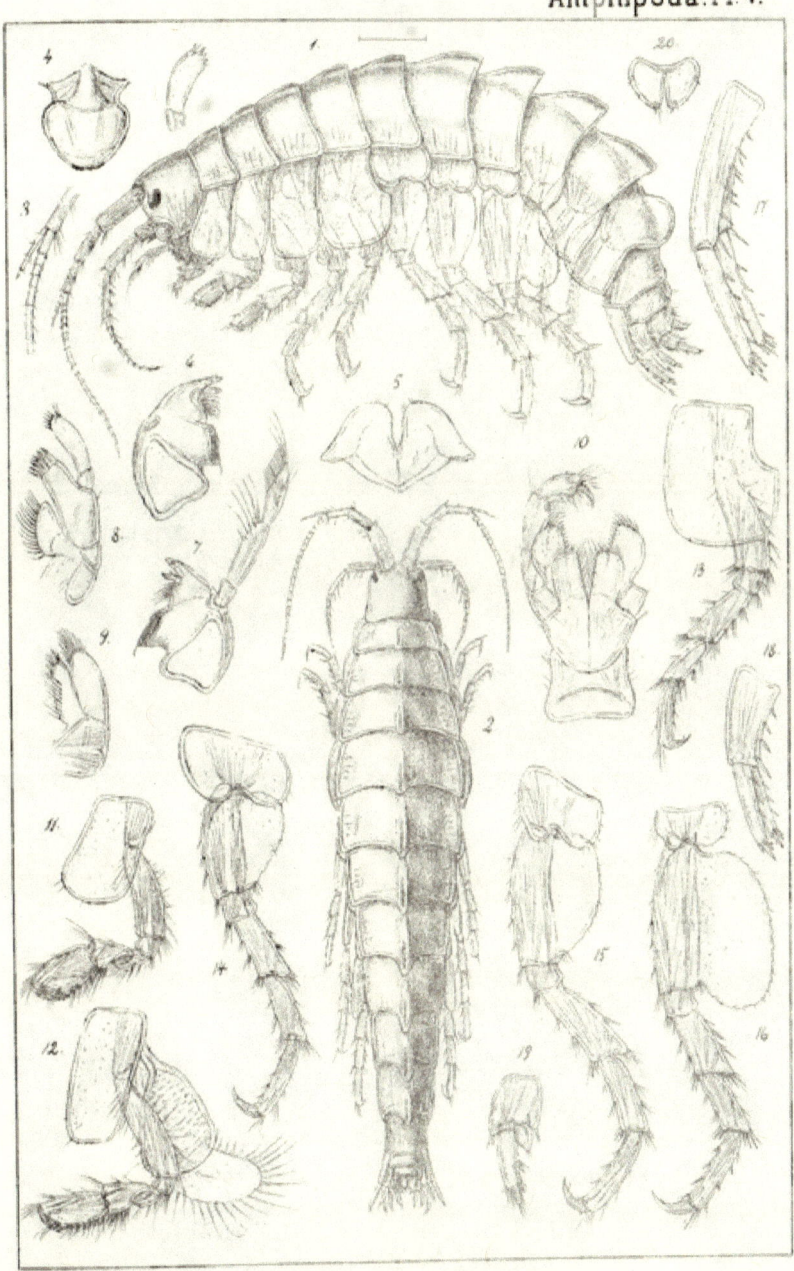

Amathillina cristata, Grimm.

G.O.Sars Crustacea caspia.
Amphipoda. Pl. VI.

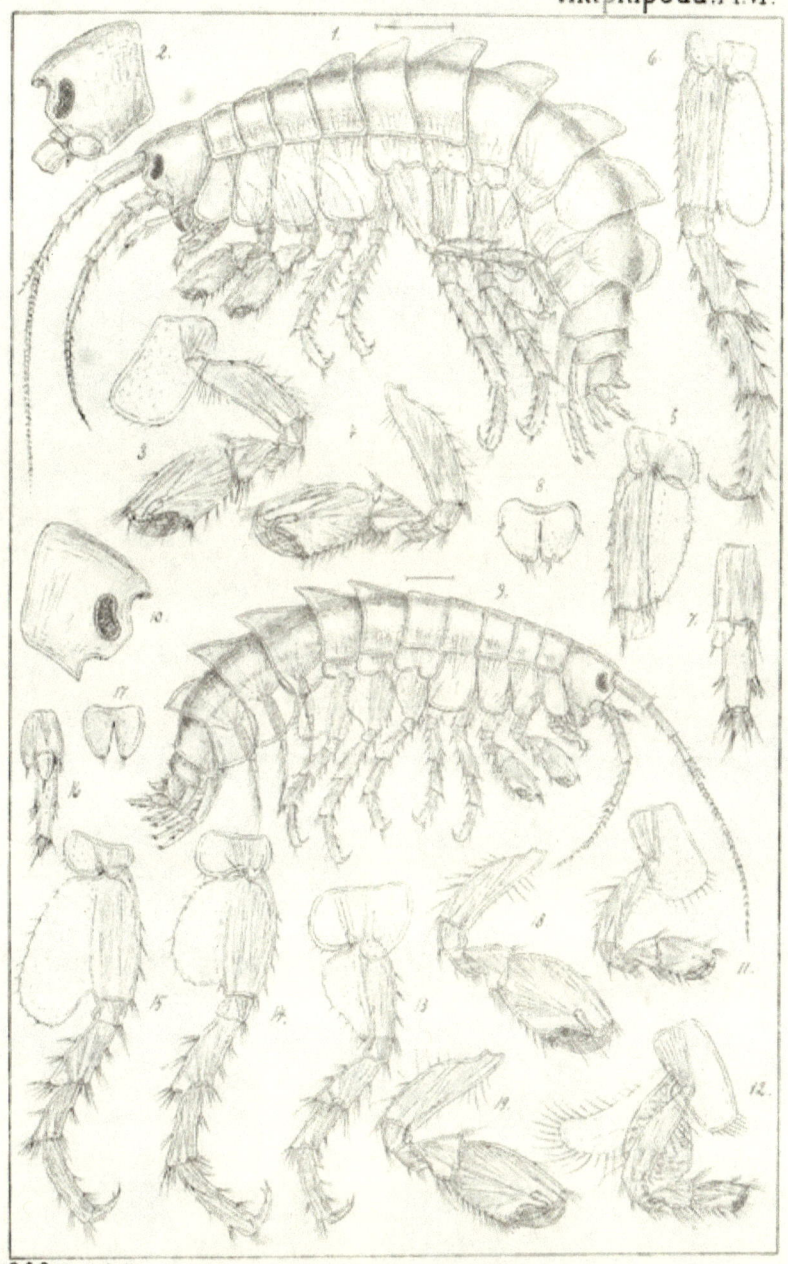

G.O.Sars autogr.

Figs. 1-8. Amathillina cristata, Grimm, (contin.)
Figs. 9-19 Amathillina affinis, n. sp.

G.O.Sars Crustacea caspia.
Amphipoda. Pl. VII.

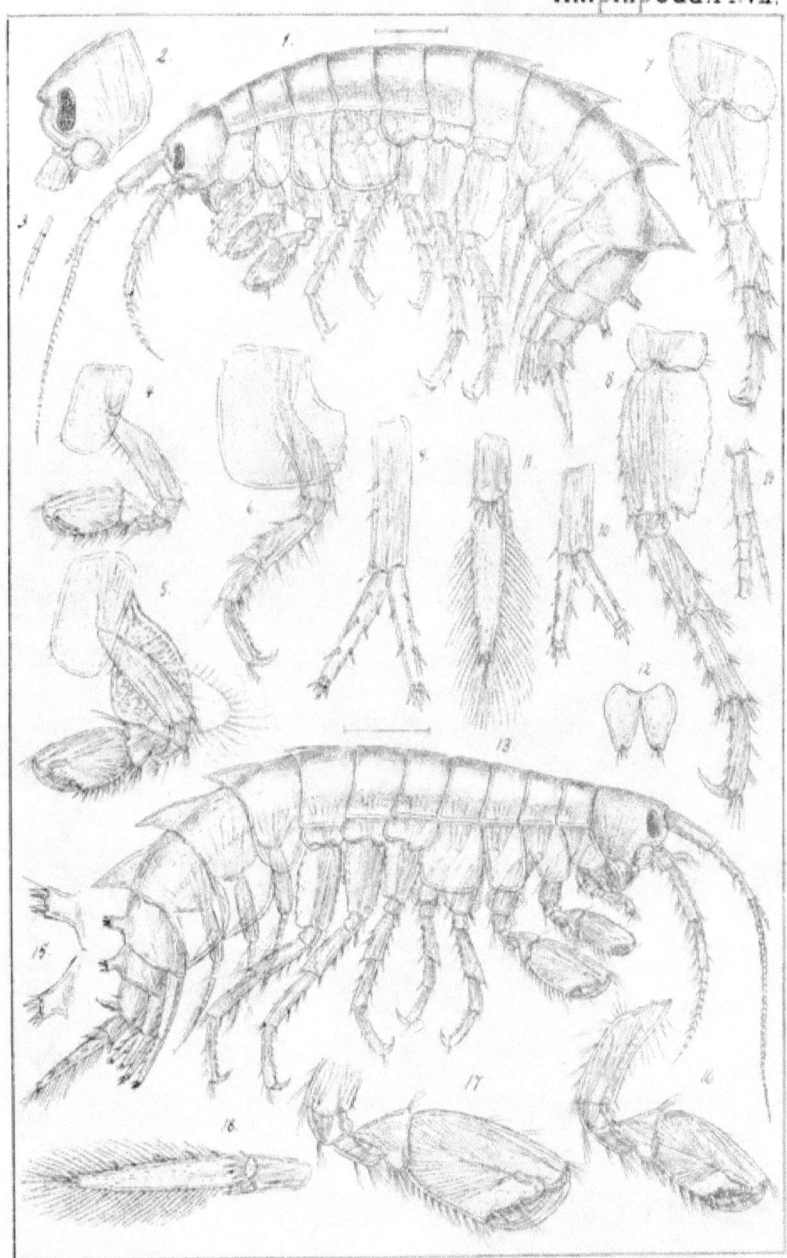

G.O.Sars autogr.
Gammarus caspius, Pallas.

G.O.Sars Crustacea caspia.
Amphipoda. Pl. VIII.

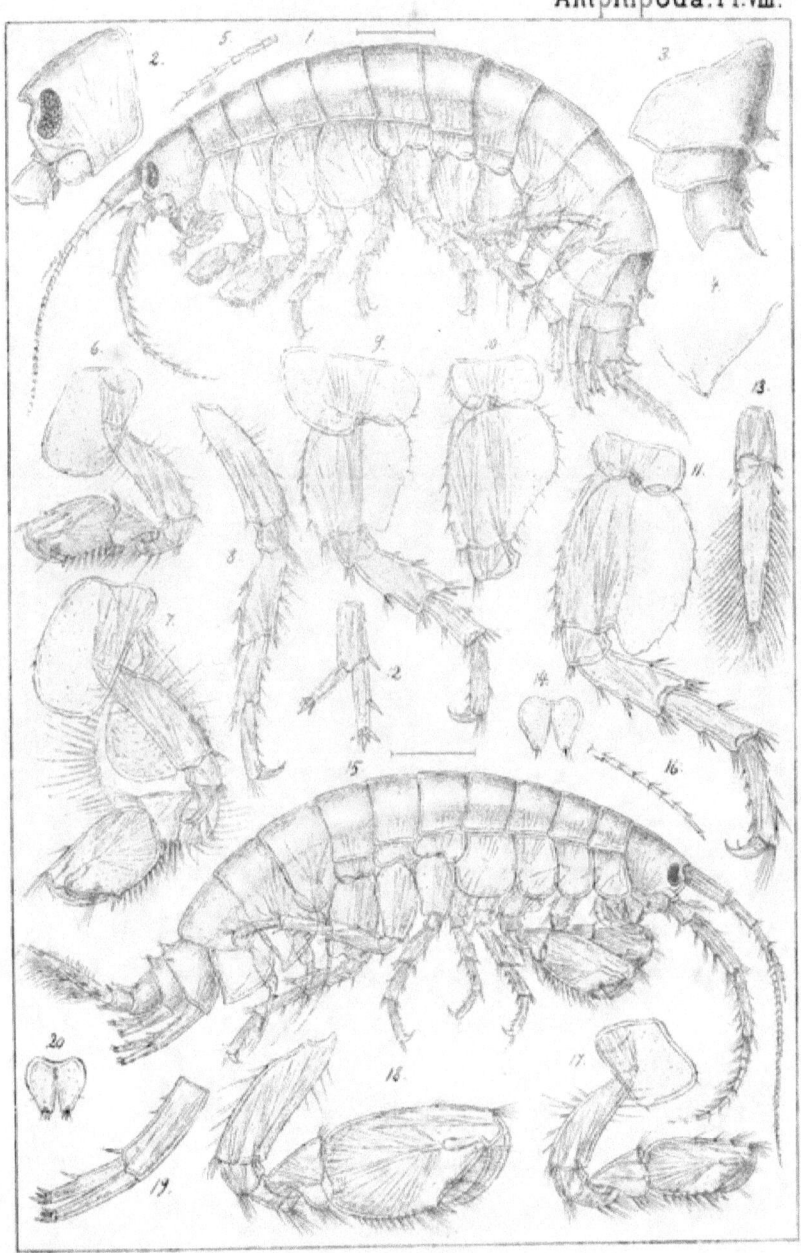

G.O.Sars autogr.

Gammarus hæmobaphes, Eichwald.

ОГЛАВЛЕНІЕ. — SOMMAIRE

	Стр.
Извлеченія изъ протоколовъ засѣданій Академіи.	27
С. Коржинскій. Замѣтка о нѣкоторыхъ видахъ Jurinea.	113
В. Серафимовъ. Наблюденія надъ пятнами на дискѣ Юпитера (съ двумя таблицами рис.).	131
С. Костинскій. О параллаксѣ звѣзды β Кассіопеи.	155
Ѳ. Бредихинъ. Изодинамы и синхроны кометы 1893 г. IV (съ 1 табл. рис.).	165
Г. О. Сарсъ. Каспійскія ракообразныя. Матеріалы для изученія карцинологической фауны Каспійскаго моря (съ 8 таблицами рис.).	178

	Pag.
Extraits de procès verbaux des séances de l'Académie.	27
S. Korshinsky. Note sur quelques espèces de Jurinea.	113
W. Séraphimof. Observations des taches sur le disque de Jupiter (avec deux planches).	131
S. Kostinsky. Sur la parallaxe de β Cassiopeiae.	155
Th. Brédikhine. Les isodynames et les synchrones de la comète 1893 IV (avec une planche).	165
G. O. Sars. Crustacea caspia. Contributions to the knowledge of the Carcinological Fauna of the Caspian Sea (with 8 autographic plates).	178

Напечатано по распоряженію Императорской Академіи Наукъ.
Сентябрь 1894 г.   Непремѣнный секретарь, Академикъ *Н. Дубровинъ*.

ТИПОГРАФІЯ ИМПЕРАТОРСКОЙ АКАДЕМІИ НАУКЪ.
Вас. Остр., 9 линія, № 12.

ИЗВѢСТІЯ

ИМПЕРАТОРСКОЙ АКАДЕМІИ НАУКЪ.

ТОМЪ I. № 4.

1894. ДЕКАБРЬ.

BULLETIN

DE

L'ACADÉMIE IMPÉRIALE DES SCIENCES

DE

ST.-PÉTERSBOURG.

Vᵉ SÉRIE. VOLUME I. № 4.

1894. DÉCEMBRE.

С.-ПЕТЕРБУРГЪ. 1894. ST.-PÉTERSBOURG.

Продается у коммиссіонеровъ Императорской Академіи Наукъ:	Commissionnaires de l'Académie Impériale des Sciences:
И. Глазунова, М. Эггерса и Комп. и К. Л. Риккера въ С.-Петербургѣ.	MM. J. Glasounof, Eggers & Cie. et C. Ricker à St.-Pétersbourg.
Н. Киммеля въ Ригѣ.	M. N. Kymmel à Riga.
Фосса (Г. Гэссель) въ Лейпцигѣ.	Voss' Sortiment (G. Haessel) à Leipzig.

Цѣна: 1 р. — Prix: 2 Mk. 50 Pf.

Crustacea caspia.
Contributions to the knowledge of the Carcinological Fauna of the Caspian Sea.
By G. O. Sars,
Professor of Zoology at the University of Christiania, Norway.

Part III.
AMPHIPODA.
Second Article:
Gammaridæ (continued);
with 8 autographic plates.

(Lu le 28 septembre 1894).

8. Gammarus Warpachowskyi, G. O. Sars, n. sp.
(Pl. IX).

Specific Characters. — Body comparatively slender and compressed, with the back perfectly smooth. Lateral lobes of cephalon forming in front a distinct angle, and being defined behind by a very slight emargination. Anterior pairs of coxal plates of moderate size, and edged distally with short, scattered bristles; 1st pair scarcely expanded in their outer part; 4th pair not as broad as they are deep. The last 2 pairs of epimeral plates of metasome slightly produced at the lateral corners. Urosome without any dorsal projections, but having on each of the segments dorsally a fascicle of delicate hairs, from among which, in the 2nd segment, rise 2 small spinules, the last 2 segments being moreover armed on each side of the dorsal face with 2 or 3 juxtaposed spines. Eyes of moderate size, and oblong oval in form. Superior antennæ scarcely exceeding in length $\frac{1}{3}$ of the body, joints of the peduncle successively diminishing in size, flagellum somewhat longer than the peduncle, accessory appendage small, biarticulate. Inferior antennæ shorter than the superior, and in male provided with a very conspicuous clothing of extremely delicate and slender sensory hairs. Gnathopoda in female comparatively feeble, in male much stronger and nearly equal-sized, with the propodos subclavate in form. Anterior pairs of pereiopoda rather narrow; the 3 posterior pairs comparatively strongly built and but little elongated, with the carpal joint rather short; basal joint of last pair slightly expanded, with the posterior edge almost angularly bent below the middle. Last pair of uropoda scarcely as long as the urosome, inner ramus very small, outer one somewhat flattened, with several fascicles of spines, but without ciliated setæ, terminal joint small. Telson short and broad, nearly semicircular in outline, each half armed with 3 slender lateral spines and a single apical one, cleft very narrow. Length of adult female 6 mm., of male 7 mm.

Remarks. — The present new species, which I have much pleasure in dedicating to the diligent explorer of the North Caspian Sea, Mr. Warpachowsky, is easily distinguishable from the other species here described by the angularly produced lateral lobes of the cephalon, the dense hairy clothing of the inferior antennæ in the male, and the structure of the gnathopoda in that sex, and finally, by the comparatively short and stout posterior pairs of pereiopoda, and more particularly the peculiar shape of the basal joint in the last pair.

Description of the female.

The length of fully adult, ovigerous specimens scarcely exceeds 6 mm., and this form accordingly belongs to the smaller species of the genus.

The form of the body (see fig. 1) is rather slender and compressed, and the back perfectly smooth, without any trace of dorsal projections.

The cephalon (fig. 2) about equals in length the first 2 segments of mesosome combined, and is but very slightly produced in front. The lateral lobes are not very prominent, and terminate anteriorly in an acutangular corner, they being defined behind by a very slight emargination.

The anterior pairs of coxal plates are of moderate size, and fringed on their distal edge with scattered bristles. The 1st pair (see fig. 5) are scarcely expanded distally, being obtusely truncated at the tip. The 3 succeeding pairs (see fig. 6) are but little broader, and have the extremity somewhat obliquely truncated. The 4th pair (see fig. 7) are, as usual, the largest, though not nearly as broad as they are deep, and have the posterior expansion vertically truncated and edged with 4 bristles.

The 3 posterior pairs of coxal plates (see figs. 8—10) exhibit the usual shape.

The epimeral plates of the metasome are well developed, the 1st pair being rounded, whereas the other 2 have the lateral corners slightly produced.

The urosome (comp. fig. 19) does not exhibit any distinct dorsal projections; but each segment has, in the middle of the dorsal face, a fascicle of fine hairs. In the 2nd segment 2 small spinules are found among the hairs, and this segment has moreover on each side of the dorsal face an obliquely transverse row of 3 somewhat stronger spines accompanied by 2 or 3 small hairs. In the last segment occurs a similar row of lateral spines, but their number is here only 2 on each side.

The eyes (see fig. 2) are of moderate size and oblong oval in form, being placed close to the anterior edges of the cephalon, and extending below nearly to the inferior edge of the lateral lobes. They have the visual elements well developed and the pigment dark.

The superior antennæ (fig. 3) are not very much elongated, scarcely exceeding in length $^1/_3$ of the body, and but sparingly supplied with short bristles. The joints of the peduncle successively diminish in size, the last one being about half the length of the 1st. The flagellum somewhat exceeds the peduncle in length, and is composed of only 9 articulations. The accessory appendage is very small, considerably shorter than the last peduncular joint, and is composed of only 2 articulations, the last of which is extremely minute.

The inferior antennæ (fig. 4) are considerably shorter than the superior, and have the last 2 joints of the peduncle nearly of same length, and provided with scattered fascicles of slender bristles. The flagellum is about half the length of the peduncle, and composed of 5 articulations.

The anterior gnathopoda (fig. 5) are comparatively small, with the carpus rather short and expanded below to a rounded setiferous lobe. The propodos is oval quadrangular in form, with about 3 fascicles of bristles below, and a single one above, near the tip. The palm is short and almost transverse, being defined below by an obtuse angle carrying a few slender spines and several bristles. The dactylus is not very strong.

The posterior gnathopoda (fig. 6) are not at all stronger than the anterior, but a little more slender, with the carpus somewhat larger, and the propodos longer in proportion to its breadth.

The 2 anterior pairs of pereiopoda (fig. 7) are rather narrow and edged with fascicles of slender spines.

The 3 posterior pairs of pereiopoda (figs. 8—10) are on the whole comparatively short and stout, and have their outer part edged with fascicles of slender spines. The last 2 pairs are about same length, whereas the antepenultimate pair are, as usual, somewhat shorter. In all pairs the carpal joint is comparatively short and thick, not nearly attaining to the length of the propodal one. The basal joint of the antepenultimate pair (fig. 8) is subquadrangular in form, with the posterior edge nearly straight, and the infero-posteal corner somewhat produced; that of the penultimate pair (fig. 9) is considerably broader in its proximal part than at the end, being expanded above to a rounded lobe edged with about 4 bristles. The basal joint of the last pair (fig. 10) is considerably broader than that of the 2 preceding pairs, being greatly expanded posteriorly, with the edge of the expansion fringed with several strong bristles, and almost angularly bent below the middle.

The 2 anterior pairs of uropoda (figs. 11, 12) have the rami subequal and linear in form, each being tipped by a number of spines, one of which is more elongated than the others. The inner ramus has besides a small lateral spine about in the middle of the upper edge; otherwise the rami are quite smooth.

The last pair of uropoda (fig. 13) do not attain the length of the urosome, and have the inner ramus very small, with a minute apical spine accompanied by 2 small hairs. The outer ramus is comparatively broad and flattened, though tapering distally. It is devoid of ciliated setæ, but has on the outer edge 3 fascicles of spines accompanied by a few simple bristles, and on the inner edge 2 similar fascicles. The terminal joint is rather small and narrow conical in form, being surrounded by several spines and bristles issuing from the end of the proximal joint.

The telson (fig. 14) is comparatively short and broad, nearly semicircular in outline, each half being armed with 3 slender lateral spines and a single apical one accompanied by 2 small hairs. The cleft is very narrow, and as usual extends to the very base of the telson.

The adult male (fig. 15) is a little larger than the female, attaining a length of 7 mm.

In the general form of the body it does not differ much from the female, though, as usual, somewhat more slender and compressed, and having the coxal plates shallower.

The superior antennæ are nearly of same appearance as in the female, whereas the inferior ones (fig. 16) are very markedly distinguished by a dense clothing of extremely delicate and slender sensory bristles, arranged in several fascicles along the posterior edge of both the peduncle and the flagellum, giving these organs a brush-like appearance.

The gnathopoda (figs. 17, 18) are much more strongly built than in the female, and nearly equal both in size and structure, exhibiting an aspect rather similar to that met with in the males of the genera *Gmelina* and *Amathillina*. As in those genera, the propodos in both pairs is very large and almost clavate in shape, with the palm much shorter than the hind margin and slightly concave, being defined below by a somewhat projecting corner armed with 2 strong spines. The dactylus is rather strong and curved, and impinges, when closed, with the tip inside the inferior corner of the propodos.

The pereiopoda exhibit exactly the same structure as in the female.

The last pair of uropoda are perhaps a little larger, but otherwise do not differ in their structure from those in the female.

Occurrence. — This form has been collected by Mr. Warpachowsky in 9 different Stations of the North Caspian Sea. Of these, one (St. 50) is located off the Tschistyi-Bank, another (St. 61) far north, outside the Bai Bogatuj Kultuk, 2 others (St. 54, 55) at some distance north of the islands Kulaly and Morskoy, the remaining Stations (17, 21, 27, 29, 52) distributed over the tract north of the peninsula Mangyschlak. In none of the Stations did it occur in any abundance.

In the collection of Dr. Grimm, this form is represented by a few specimens collected in the Bai of Baku, partly in quite shallow water, among grass, partly from a depth of 2—3 fathoms.

9. Gammarus minutus, G. O. Sars, n. sp.

(Pl. X, figs. 1—26).

Specific Characters. — Body comparatively short and stout, with evenly rounded back. Lateral lobes of cephalon somewhat projecting and broadly rounded at the tip. Anterior pairs of coxal plates of moderate size and rather densely setous at the distal edge; 1st pair very slightly widening distally; 4th pair not so broad as they are deep. The last 2 pairs of epimeral plates of metasome nearly rectangular. Urosome comparatively short and somewhat gibbous above, the 1st segment partly overlapping the 2nd dorsally, both having a very small dorsal fascicle of hairs; last segment with a single spinule on each side. Eyes not very large, and oblong oval in form. Superior antennæ but little longer than the inferior, and having the basal joint rather large, flagellum about the length of the peduncle, accessory appendage 3-articulate. Gnathopoda in female comparatively small and nearly equal-sized; those in male much more powerful and rather unequal, propodos of the anterior ones obpyriform, of the posterior ones very large and oval quadrangular in form. Anterior pairs of pereiopoda rather densely setous, and having the meral joint somewhat expanded; the 3 posterior pairs moderately slender; basal joint of last pair very large, forming posteriorly a broadly rounded expansion somewhat projecting at the infero-posteal corner. Last pair of uropoda with the inner ramus very small, outer one sublinear, with a single fascicle of spines about in the middle of the outer edge; terminal joint small. Telson with the lateral lobes comparatively narrow and tipped with 2 spines, cleft rather wide. Length of adult female 4 mm., of male 5 mm.

Remarks. — This is a very small species, indeed the smallest as yet known, and is also easily distinguished from the preceding species by the comparatively stout body, the short and, as it were, gibbous urosome, and by the structure of the gnathopoda and pereiopoda.

Description of the female.

The length of fully adult, ovigerous specimens scarcely exceeds 4 mm.

The form of the body (see fig. 1) appears on the whole rather short and stout, being not nearly so much compressed as in the preceding species. As in the latter, the back is quite smooth throughout, and broadly rounded.

The cephalon about equals in length the first 2 segments of mesosome combined, and has the lateral lobes somewhat projecting and broadly rounded at the tip, being defined from the postantennal corners by a rather deep emargination, encircling the large, globular basal joint of the inferior antennæ.

The anterior pairs of coxal plates are nearly twice as deep as the corresponding segments, and have their distal edge densely fringed with delicate bristles. The 1st pair (see fig. 4) are slightly expanded in their outer part, whereas the 2 succeeding pairs (see fig. 5) are almost of equal breadth throughout. The 4th pair (see fig. 6) are, as usual, larger than the preceding pairs, being somewhat expanded below the posterior emargination, though not nearly as broad as they are deep.

The epimeral plates of the metasome are well developed, the 1st pair being, as usual, the smallest, and evenly rounded, whereas the 2 posterior pairs terminate in an angle.

The urosome (see fig. 1) is comparatively short and stout, without any dorsal projections, but having the dorsal face of the first 2 segments strongly convex, and as it were gibbous. The 2nd segment is very short in its dorsal part, being to some extent overlapped by the 1st, and, like the latter, has dorsally a small fascicle of hairs, whereas lateral spinules are wholly absent. The last segment has on either side a single small spinule, but no dorsal fascicle.

The eyes are not very large, but of an oblong oval form, with well-developed visual elements and dark pigment.

The superior antennæ (fig. 2) are comparatively short, but little exceeding in length ¼ of the body, and have the 1st joint of the peduncle rather large, nearly twice as long as the other 2 combined. The flagellum is about the length of the peduncle, and composed of only 8 articulations. The accessory appendage equals in length ¼ of the flagellum, and is composed of 3 distinctly defined articulations.

The inferior antennæ (fig. 3) are a little shorter than the superior, and have the antepenultimate joint of the peduncle rather thick, and projecting posteriorly as an angle tipped by several slender bristles. The penultimate joint forms likewise posteriorly a slight angular expansion provided with a number of slender bristles, and is somewhat longer than the last one. The flagellum is about half the length of the peduncle, and is composed of only 4 articulations.

The gnathopoda (figs. 4, 5) are rather small, and almost exactly of same shape, the propodos being in both pairs of an oval quadrangular form, with the palm rather short and almost transverse.

The 2 anterior pairs of pereiopoda (fig. 6) are rather densely supplied with slender bristles, and have the meral joint comparatively large and expanded distally.

The 3 posterior pairs of pereiopoda (figs. 7—9) are moderately slender and but little different in length, having their outer part edged with fascicles of slender bristles. The basal joint of the antepenultimate pair (fig. 7) is rather broad and subquadrangular in form, with the infero-posteal corner nearly rectangular: that of the penultimate pair (fig. 8) is considerably narrower, though somewhat expanded in its proximal part. The last pair (fig. 9) differ considerably from the others in the unusually large size of the basal joint, which forms posteriorly a very broad and evenly rounded expansion edged with a few small hairs, and projecting below as a broadly rounded lobe reaching somewhat beyond the ischial joint.

The 2 anterior pairs of uropoda (fig. 10) have the rami quite smooth except at the tip, which carries the usual fascicle of spines.

The last pair of uropoda (fig. 11) are of moderate size, with the inner ramus very small and tipped by a single minute spinule. The outer ramus is narrow linear in shape, and only provided with a single lateral fascicle of spines occurring somewhat beyond the middle of the outer edge. The terminal joint is very small, narrow conical in form, and tipped by 3 small bristles.

The telson (fig. 16) has the lateral halves rather narrow and each armed on the obtusely pointed tip, with 2 small spinules. The cleft, which, as usual, extends to the base of the telson, gradually widens distally.

The adult male (fig. 12) is a little larger than the female, attaining a length of about 5 mm.

It resembles the female in the general form of the body, but is easily recognized by the somewhat shallower coxal plates, and especially by the structure of the gnathopoda.

The latter appendages (figs. 13, 14) are much more powerfully developed than in the female, and also rather unequal in size, the posterior ones (fig. 14) being much stronger than the anterior, with the propodos very large and of an oval quadrangular form. In the anterior pair (fig. 13) the propodos is also rather large, as compared with that in the female, but much narrower than in the posterior pair, and nearly obpyriform in shape, with the palm very oblique. In both pairs the palm is defined below by a distinct, though somewhat obtuse angle armed with several strong spines.

The last pair of uropoda (fig. 15) are a little larger than in the female, about equalling in length the urosome, but are otherwise of much the same structure.

Occurrence. — Of this form, a few specimens were collected by Mr. Warpachowsky at St. 52, lying north of the island Swjatoj. This is the only place where the species has hitherto been found. In the collection of Dr. Grimm I have not yet succeeded in detecting any specimen of this species.

10. Gammarus macrurus, G. O. Sars, n. sp.

(Pl. X, figs. 17—27).

Specific Characters. — Body slender and quite smooth throughout. Lateral lobes of cephalon somewhat projecting and evenly rounded at the tip. Anterior pairs of coxal plates rather large and closely contiguous; 1st pair slightly widening distally; 4th pair very large, being fully as broad as they are deep. The last 2 pairs of epimeral plates of metasome obtusely produced at the lateral corners. Urosome rather slender and smooth above; last segment with a small spinule on either side. Eyes of moderate size and oblong oval in form. Antennae rather short and equal-sized, but little exceeding in length $1/4$ of the body, the superior ones with the 1st joint of the peduncle rather large, flagellum about the length of the peduncle, accessory appendage 3-articulate. Gnathopoda in female extremely small and feeble, with the propodos scarcely larger than the carpus. Anterior pairs of pereiopoda normally developed; the 3 posterior pairs moderately slender; basal joint of antepenultimate pair rather broad and rounded at the infero-posteal corner, that of last pair considerably expanded and of regular oval form. Last pair of uropoda considerably exceeding the urosome in length, inner ramus small, outer very much elongated, with the terminal joint well developed, being about half the length of the proximal one. Telson rather narrow, each half having 1 lateral and 1 apical spine, cleft narrow. Length of adult female 6 mm.

Remarks. — This species may be readily distinguished from the 2 preceding ones by the very slender form of body, the comparatively small and equal-sized antennæ, the extremely feeble gnathopoda, and especially by the very much elongated last pair of uropoda, which latter characteristic has given rise to the specific name. Only female specimens have hitherto come under my inspection.

Description of the female.

Fully adult, ovigerous specimens scarcely exceed 6 mm. in length, and accordingly, this species also belongs to the small-sized species of the genus.

The form of the body (see fig. 17) is very slender and compressed, with the back perfectly smooth throughout.

The cephalon scarcely attains the length of the first 2 segments of mesosome combined, and has the lateral lobes somewhat projecting and broadly rounded at the tip, being defined behind by a rather deep emargination.

The anterior pairs of coxal plates are comparatively large and closely contiguous, forming together on each side a perfectly continuous wall. Their distal edge is only fringed with very small and scattered bristles. The 1st pair (see fig. 19) are slightly expanded in their outer part, and obtusely truncated at the tip; the 2 succeeding pairs are of a more regular, oblong quadrangular form. The 4th pair (see fig. 21) are very large and greatly expanded in their outer part, being fully as broad as they are deep, and exhibiting an irregularly angular shape, with the posterior expansion vertically truncated.

The epimeral plates of the metasome are comparatively large, the last 2 being produced at the lateral corners to a somewhat obtuse point.

The urosome is rather slender and perfectly smooth above, with only a very small spinule on each side of the dorsal face in the last segment.

The eyes are of moderate size and oblong oval form, with well-developed visual elements and dark pigment.

The antennæ (see fig. 17) are unusually short and nearly equal-sized, scarcely exceeding in length ¼ of the body, and are supplied with scattered fascicles of slender bristles. The superior ones (fig. 18) have the 1st joint of the peduncle very large, being nearly twice as long as the other 2 combined. The 3rd joint is rather short, scarcely longer than it is broad. The flagellum is about the length of the peduncle, and composed of 6 articulations only. The accessory appendage about equals in length the last 2 peduncular joints combined, and is composed of 3 articulations. The inferior antennæ nearly agree in their structure with those in *G. minutus*.

The gnathopoda (figs. 19, 20) are extremely small and feeble, the posterior ones being somewhat more slender than the anterior. The propodos in both pairs is scarcely broader than the carpus and about equals it in length. In the posterior pair both the carpus and the propodos appear somewhat more elongated than in the anterior one.

The 2 anterior pairs of pereiopoda (fig. 21) are of moderate size and resemble those in *G. minutus*, except that the meral joint is somewhat less expanded.

The 3 posterior pairs of pereiopoda (figs. 22—24) are rather slender, and have their outer part supplied with fascicles of slender bristles. The basal joint of the antepenultimate pair (fig. 22) is very broad and of a rounded quadrangular form, with the infero-posteal corner rounded off; that of the penultimate pair (fig. 23) is considerably smaller, and has the posterior edge evenly curved. The last pair (fig. 24) have, as usual, the basal joint larger than that of the 2 preceding pairs and of a rather regular oval form, with the posterior edge but slightly curved, and the infero-posteal corner expanded to a rounded lobe reaching about to the end of the ischial joint. The outer part of these legs, in all the specimens, was broken off.

The 2 anterior pairs of uropoda (fig. 25) have the rami equal-sized and narrow linear in form, being edged with a few lateral spines in addition to the usual apical ones.

The last pair of uropoda (fig. 26) are remarkable by their unusual length, even considerably exceeding that of the urosome. The basal part is comparatively short, and armed at the end below with a transverse row of strong spines. The inner ramus, as in most of the Caspian species, is very small, terminating with 2 slender spines. The outer ramus, on the other hand, is unusually elongated and rather slender, with the terminal joint well developed and occupying about the third part of the length of the ramus. The proximal joint has outside 2 small fascicles of spines, and inside a row of about 6 slender, ciliated setæ; at the end it carries, moreover, a few spines and simple bristles. The terminal joint has the outer edge smooth, the inner provided with a row of 4 setæ, and moreover carries on the tip 3 slender bristles.

The telson (fig. 27) is rather narrow, being much longer than it is broad at the base, and gradually tapers distally. Each half is armed with a small lateral spinule and another apical one accompanied by a small hair. The cleft is rather narrow and extends to the base of the telson.

Occurrence. — Only a few specimens of this form were found among other Gammari collected by Mr. Warpachowsky at St. 53 and 54, both located at some distance north of the islands Kulaly and Morskoj. In the collection of Dr. Grimm I have not yet succeeded in detecting any specimen of this species.

11. Gammarus compressus, G. O. Sars, n. sp.

(Pl. XI, figs. 1—19).

Specific Characters. — Body moderately slender and very much compressed, with the back smooth throughout. Lateral lobes of cephalon but slightly projecting, and narrowly rounded at the tip. Anterior pairs of coxal plates rather large and densely fringed with bristles distally; 1st pair very much expanded in their outer part; 4th pair large, about as broad as they are deep. The last 2 pairs of epimeral plates of metasome but very slightly produced at the lateral corners. Urosome comparatively short and quite smooth above. Eyes not very large, narrow oblong in form. Antennae comparatively short and nearly equal-sized, the superior ones having the 1st joint of the peduncle very large and massive, flagellum but little longer than the peduncle, accessory appendage 3-articulate. Gnathopoda in female small and feeble, and nearly equal-sized. Anterior pairs of pereiopoda comparatively strongly built, and densely setous, with the meral joint not a little expanded, especially in the 2nd pair; the 3 posterior pairs rather slender and nearly equal in length; basal joint of last pair moderately expanded and regularly oval in form. Last pair of uropoda with the inner ramus less rudimentary than in the preceding species, being nearly half as long as the outer, the latter having the terminal joint well developed. Telson fully as long as it is broad at the base, each half with 2 apical spines. Length of adult female 7 mm.

Remarks. — This new species is chiefly characterised by its extremely compressed body, the short, equal-sized antennae, the form of the 1st and 4th pair of coxal plates, and the structure of the last pair of uropoda. As of the preceding species, only female specimens have hitherto come under my inspection.

Description of the female.

The length of fully adult, ovigerous specimens measures about 7 mm., and accordingly this form must also be reckoned among the smaller species of the genus.

The form of the body (see fig. 1) is rather slender and very much compressed, with the back narrowly rounded and smooth throughout.

The cephalon about equals in length the first 2 segments of the mesosome combined, and appears nearly transversely truncated in front, the lateral lobes projecting but slightly, and being narrowly rounded at the tip.

The anterior pairs of coxal plates are rather large and fringed on their distal edge with numerous delicate bristles. The 1st pair (see fig. 3) are of a somewhat unusual form, being very much expanded in their outer part, and forming in front a linguiform lobe advancing beneath the cephalon as far as the insertion of the inferior antennæ. The 2 succeeding pairs (see fig. 4) are of regular oblong quadrangular form and transversely truncated at the tip. The 4th pair (see fig. 5) are very large and expanded, being fully as broad as they are deep, and projecting below the posterior emargination as a distinct angle.

The epimeral plates of the metasome are of moderate size, and the 2 posterior pairs but very slightly produced at the lateral corners.

The urosome is comparatively short and stout, being quite smooth above, with only an extremely minute spinule on each side of the dorsal face of the last segment.

The eyes are not very large but of a narrow oblong form, with well developed visual elements and dark pigment.

The antennæ are comparatively short and nearly equal-sized, scarcely exceeding in length $^1/_4$ of the body. The superior ones (fig. 2) have the 1st joint of the peduncle very large and massive, considerably exceeding in length the other 2 combined. The flagellum is scarcely longer than the peduncle, and is composed of only 9 articulations. The accessory appendage does not attain the length of the last 2 peduncular joints combined, and is composed of 3 articulations. The inferior antennæ have the last joint of the peduncle smaller than the penultimate one, and the flagellum nearly as long as the peduncle and 6-articulate.

The gnathopoda (figs. 3, 4) are comparatively small and feeble, resembling in structure those in the 3 preceding species. The propodos in the anterior pair (fig. 3) is a little broader than in the posterior (fig. 4), where it is scarcely larger than the carpus. In both pairs the palm is very short and nearly transverse.

The 2 anterior pairs of pereiopoda are rather strongly built and, especially the 2nd pair (fig. 5), very densely clothed with slender bristles, which form a dense fringe along the posterior edge of the meral and carpal joints. The former joint is very large and expanded, and the latter likewise unusually broad and regularly oval in form.

The 3 posterior pairs of pereiopoda (figs. 6—8) are comparatively slender and nearly of equal length, with their outer part edged with fascicles of delicate bristles. In all pairs the meral joint is somewhat expanded, whereas the carpal and propodal ones are linear in form and about equal in length. The basal joint of the antepenultimate pair (fig. 6) is of a rounded oval form, with the infero-posteal corner somewhat produced and rounded off; that of the penultimate pair (fig. 7) is considerably narrower and obliquely oblong, being expanded at the upper part posteriorly to a rounded, minutely setiferous lobe. The last pair (fig. 8) have the basal joint considerably larger than that of the 2 preceding pairs, and of a regular oval form, with the posterior edge slightly curved and fringed with small bristles, the infero-posteal corner being produced to a broadly rounded lobe extending to the end of the ischial joint.

The 2 anterior pairs of uropoda have the rami subequal and of a narrow linear form, being spinous only at the tip.

The last pair of uropoda (fig. 9) are of moderate size, and have the basal part, as usual, armed at the end below with a transverse row of strong spines. The inner ramus is less rudimentary than in most other Caspian species, being about half the length of the outer. It is conically tapering and carries at the tip a slender spine and 2 or 3 delicate bristles. The outer ramus is about twice as long as the basal part, and has the terminal joint well developed, about half the length of the proximal one. The latter has outside 2 small fascicles of spines and inside about 4 slender setae.

The telson (fig. 10) is fully as long as it is broad at the base, and is, as usual, cleft to the base, the cleft widening gradually distally. The lateral lobes are comparatively narrow, and each carry at the somewhat truncated tip 2 slender spines.

Occurrence. — Some specimens of this form were collected by Mr. Warpachowsky at St. 2, located off the Tschistyi-Bank; a single specimen was moreover found at St. 16, lying east of the island Swjatoj.

In the collection of Dr. Grimm this species seems not to be represented.

12. Gammarus similis, G. O. Sars. n. sp.

(Pl. XI, figs. 11—20).

Specific Characters. — Very like the preceding species, though not nearly so much compressed. Lateral lobes of cephalon somewhat projecting and obtusely rounded at the tip. Anterior pairs of coxal plates densely setons at the terminal edge; 1st pair but very slightly widening distally; 4th pair not so broad as they are deep. Epimeral plates of metasome about as in *G. compressus*. Urosome somewhat more slender than in that species and smooth above, last segment with 2 small spinules on each side of the dorsal face. Eyes oblong oval in form, with dark pigment. Antennae comparatively short and nearly equal-sized, the superior ones with the 1st joint of the peduncle very large, flagellum somewhat longer than the peduncle, accessory appendage 4—5-articulate. Gnathopoda in female nearly as in *G. compressus;* those in male somewhat stronger and rather unequal, the posterior ones being the larger, palm in both pairs more oblique than in the female, and having in the middle, outside, a strong spine. Pereiopoda of a similar structure to that in *G. compressus*, except that the basal joint of the last pair is somewhat more expanded. Last pair of uropoda with the inner ramus rather small, the outer elongated and having the terminal joint poorly developed. Telson not as long as it is broad at the base, each half with 2 small apical spines. Length of adult female 9 mm., of male about the same.

Remarks. — The present species is nearly allied to *G. compressus*, but on a closer examination is easily distinguishable by its far less compressed body, the different form of the 1st and 4th pair of coxal plates, and the likewise rather different structure of the last pair of uropoda. It also bears some resemblance to the form described by Mr. Sowinsky from the Azow Sea, as *G. macoticus*, which latter species also occurs in the Caspian Sea, and is represented by several well marked specimens in the collection of Dr. Grimm.

Description of the female.

The length of adult, ovigerous specimens amounts to 9 mm., and this form accordingly grows to a somewhat larger size than the preceding species.

The form of the body (see fig. 11) is, as in the latter, rather slender, but considerably less compressed, the back being broadly vaulted and quite smooth throughout.

The cephalon does not fully attain the length of the first 2 segments of mesosome combined, but has the lateral lobes rather prominent and obtusely rounded at the tip, being defined behind by a rather deep emargination.

The anterior pairs of coxal plates are of moderate size and densely setiferous on their distal edge. The 1st pair (see fig. 13) differ not a little in their shape from those in *G. compressus*, being only very slightly expanded distally, and nearly transversely truncated at the tip. Also the 4th pair (see fig. 15) are markedly distinguished in being far less expanded in their outer part.

The epimeral plates of the metasome are nearly of same appearance as in the above-mentioned species.

The urosome, on the other hand, appears considerably more elongated and slender than in that species, and has on each side of the last segment 2 minute spinules.

The eyes are comparatively a little larger than in *G. compressus* and of a more pronounced oblong reniform shape.

The antennæ, as in that species, are comparatively short and equal-sized, scarcely exceeding in length ¹/₄ of the body. The superior ones (fig. 12) have the 1st joint of the peduncle very large, nearly twice as long as the other 2 combined. The flagellum somewhat exceeds the peduncle in length, and is composed of about 11 articulations. The accessory appendage is comparatively more fully developed than in *G. compressus*, and composed of 4 to 5 articulations. The inferior antennæ have the outer joints of the peduncle rather richly supplied with bristles. The flagellum nearly attains the length of the peduncle, and is composed of about 7 articulations.

The gnathopoda (figs. 13, 14) resemble those in *G. compressus*, though the posterior ones are perhaps still more slender than in that species.

The pereiopoda (figs. 15—18) also exhibit a very similar structure to that in the above-named species, and need not therefore be described in detail. In the last pair (fig. 18), however, the basal joint is comparatively larger and more expanded, being nearly as broad as it is long.

The last pair of uropoda (fig. 18) are rather elongated, about equalling the urosome in length, and differ very markedly in their structure from those in *G. compressus*. Thus the inner ramus is much smaller, being scarcely ¹/₃ as long as the outer, and the terminal joint of the latter is likewise very minute, as compared with that in the said species. The proximal joint of the latter ramus is, on the other hand, much elongated and of a linear form, with several slender setæ on the inner edge and 2 or 3 small fascicles of spines on the outer.

The telson (fig. 20) is comparatively shorter than in *G. compressus*, being not nearly so long as it is broad at the base; otherwise it exhibits a very similar structure.

The adult male is about same size as the female, and does not greatly differ from it in outward appearance. It is, however, easily recognized by the somewhat more strongly built gnathopoda. As in most other Gammari, the posterior pair (fig. 22) are more powerful than the anterior (fig. 21), the propodos being in the former considerably larger and broader. In both pairs the palm is somewhat oblique, and armed in the middle, outside, with a strong spine in addition to those occurring on the lower corner.

Occurrence. — This species has been collected by Mr. Warpachowsky in 6 different Stations of the North Caspian Sea. Of these, one (St. 16) is located off the island Swjatoj; another (St. 21) at the northern point of the peninsula Mangyschlack, 2 others (St. 53, 54) at some distance north of the islands Kulaly and Morskoj, the last 2 (St. 61, 63) in the northern and eastern part of that Sea.

In the collection of Dr. Grimm, several specimens of this species are to be found, partly collected at Baku from stones on the shore, partly on the west coast of Sara, among *Zostera*.

13. Gammarus robustoides, Grimm MS.

(Pl. XII).

Syn.: ? *Gammarus caspius*, Sp. Bate (not Pallas)
Gammarus aralo-caspius, Grimm MS.

Specific Characters. — Body rather strongly built and but little compressed, the back being broadly rounded. Lateral lobes of cephalon rather projecting and somewhat obliquely truncated at the tip, the lower corner being more prominent than the upper. Anterior pairs of coxal plates of moderate size, and densely fringed with bristles on their distal edge; 1st pair but very slightly widening distally; 4th pair scarcely as broad as they are deep. The last 2 pairs of epimeral plates of metasome acutely produced at the lateral corners. Urosome with the 2 anterior segments forming each at the end dorsally a slight protuberance armed with a number of densely crowded spines arranged in an angularly bent transverse row, the 1st segment having generally 8 such spines, the 2nd 6; last segment with a dorsal fascicle of delicate hairs, and having besides on either side of the dorsal face 2 spinules. Eyes of moderate size and oval reniform in shape. Antennæ comparatively short and nearly equal-sized, the superior ones with the 1st joint of the peduncle rather large, flagellum somewhat exceeding the peduncle in length and composed of numerous short articulations, accessory appendage well developed, 6—8-articulate. Gnathopoda in female moderately strong and somewhat unequal, the posterior ones being the larger, carpus in both pairs rather short, propodos well developed, with the palm somewhat oblique; those in male considerably more strongly built than in female, with the propodos, especially of the posterior ones, very large and tumid. Anterior pairs of pereiopoda rather robust, and very densely setiferous, with the meral joint large and expanded; the 3 posterior pairs moderately slender, and having their outer part clothed with numerous fascicles of bristles and scattered spines; basal joint of last pair oblong oval in form, being somewhat broader in female than in male, hind edge densely fringed with short bristles, infero-posteal corner produced to a short, narrowly rounded lobe. Last pair of uropoda of moderate size, with the inner ramus very small, outer one well developed and densely fringed with slender, partly ciliated setæ, terminal joint extremely minute. Telson about as long as it is broad, each half with 3 or 4 apical spines. Length of adult female 17 mm., of male reaching 22 mm.

Remarks. — The *Gammarus caspius* of Sp. Bate may perhaps be referable to this species, but, as above stated, this name was given a long time ago by Pallas to a very different form. In the collection of Dr. Grimm this species has been labelled in 2 different manners. On one bottle, containing an unusually large male specimen, in which, by some accident, the dorsal spines of the 1st segment of the urosome were rubbed off, the name *G. robustoides* is given; 2 other bottles, containing several considerably smaller specimens of the same species, are labelled with the name *G. aralo-caspius*. I think I am right in preferring the former name, as the latter is inconveniently near *G. caspius*, which is a very different form. The species is easily recognized by the strong and densely crowded dorsal spines occurring on the 2 anterior segments of the urosome.

Description of the female.

The length of adult ovigerous specimens amounts to about 17 mm., but in some places they would seem not to attain to such a large size. In every case this form must be reckoned among the larger-sized species of the genus.

The body (see fig. 1) is of a rather robust form and but little compressed, the back being broadly rounded and perfectly smooth.

The cephalon about equals in length the first 2 segments of mesosome combined, and has the lateral lobes rather projecting and somewhat obliquely truncated at the tip, with the lower corner more prominent than the upper. They are defined posteriorly by a rather deep emargination encircling the large, globular basal joint of the inferior antennæ.

The anterior pairs of coxal plates are of moderate size, and densely fringed on their distal edge with delicate bristles. The 1st pair (see fig. 3) are very slightly expanded in their outer part, whereas the 2 succeeding pairs (see fig. 4) are nearly of same breadth throughout. The 4th pair (see fig. 5) are, as usual, the largest, though scarcely as broad as they are deep; their posterior expansion is vertically truncated, and, like the distal edge, densely fringed with bristles.

The epimeral plates of the metasome are well developed, and the 2-nd pair a little deeper than the last, both being acutely produced at the lateral corners.

The urosome (comp. figs. 17, 18) is of moderate size, and has the 2 anterior segments somewhat elevated at the end dorsally, whereby 2 obtuse dorsal prominences are formed, each armed with a number of densely crowded spines arranged in a somewhat angularly bent transverse row. The number of these spines is generally on the 1st segment 8, on the 2nd 6. In the last segment occurs a dorsal fascicle of delicate hairs, and, in addition, on either side of the dorsal face 2 spinules.

The eyes are of an oval reniform shape, with well-developed visual elements and dark pigment. In some specimens, however, preserved from older time in the Museum of St. Petersburg, and collected at Baku, no trace of any ocular pigment was observable; but whether this was merely due to the action of the spirit, or to some different nature of the pigment in those specimens, I am unable to state.

The antennæ are comparatively short and but little different in length, differing in this respect from what is the case in the typical Gammari. The superior ones (fig. 2) but slightly exceed in length $\frac{1}{4}$ of the body, and have the 1st joint of the peduncle rather large and longer than the other 2 combined. The flagellum somewhat exceeds the peduncle in length, and is very flexible, being composed of numerous short articulations, amounting to 20—24 in all. The accessory appendage is well developed, though scarcely as long as the last 2 peduncular joints combined, and is composed of from 6 to 8 articulations. The inferior antennæ are a little shorter than the superior, and are more densely setiferous. The penultimate joint of the peduncle is somewhat larger than the last one, and the flagellum about equals in length those joints combined, being composed of 8—10 articulations.

The gnathopoda (figs. 3, 4) are moderately strong and somewhat unequal, the posterior ones (fig. 4) being considerably more powerful than the anterior (fig. 3). In both pairs the carpus is quite short, triangular, and expanded below to a narrow, setiferous lobe. The propodos is comparatively large and tumid, especially in the posterior pair, and has the palm somewhat oblique, being defined below by an obtuse angle armed with several strong spines.

The 2 anterior pairs of pereiopoda (fig. 5) are unusually robust and densely edged with slender bristles, especially along the posterior edges of the meral and carpal joints. The former joint is very large and expanded, and also the carpal joint gradually expands somewhat distally, whereas the propodal joint is of the usual narrow linear form.

The posterior pairs of pereiopoda (figs. 6—8) are considerably more slender, and have their outer part edged with numerous fascicles of delicate bristles, and by a number of scattered spines. The antepenultimate pair (fig. 6) are considerably shorter than the other 2, and have the basal joint rounded quadrangular in form, with the anterior edge somewhat curved, and edged with several fascicles of slender bristles, as also with a number of small spines; the infero-posteal corner of this joint is somewhat projecting and angular. In the penultimate pair (fig. 7) the basal joint is somewhat more elongated and slightly expanded in its proximal part, gradually tapering distally. The last pair (fig. 8) are about same length as the penultimate, but are markedly distinguished by the much larger size of the basal joint. This exhibits a rather regular oval form, with the posterior edge evenly curved, and, as in the preceding pairs, densely fringed throughout with comparatively short bristles; its infero-posteal corner projects below as a narrowly rounded lobe, reaching about to the end of the ischial joint. Of the outer joints, the carpal one in all 3 pairs is rather elongated and slender, exceeding in length both the meral and the propodal joints. The dactylus is not very strong, and has near the tip a small bristle.

The anterior pairs of uropoda (figs. 9 and 16) have the rami sublinear in form and armed with scattered lateral spines, their tip carrying only a single spine accompanied by 2 small denticles.

The last pair of uropoda (fig. 10) are of moderate size, and, as in most other Caspian Gammari, have the inner ramus very small and scale-like, with a single minute apical spinule and several slender bristles on the inner edge. The outer ramus is about twice the length of the basal part and slightly tapers distally. It is fringed all round with numerous slender, partly ciliated setæ, and has besides, on the outer edge, 3 fascicles of spines; the terminal joint is extremely minute.

The telson (fig. 11) is about as long as it is broad, and is, as usual, divided by a deep cleft into 2 halves, each of which carries on the narrowly truncated tip 3 or 4 spines, but no trace of any lateral ones.

The adult male (fig. 12) generally attains a considerably larger size than the female, its length amounting to 22 mm.

It does not differ conspicuously in its general form from the female, but is easily recognizable by the much stronger development of the gnathopoda.

As in the female, these limbs (figs. 14, 15) are somewhat unequal, the posterior ones (fig. 15) being considerably stronger than the anterior (fig. 14). The propodos in both pairs, but especially in the posterior ones, is very large and tumid, though of a similar shape to that in the female.

Of the other appendages, the last pair of perciopoda somewhat differ in the basal joint being narrower and less expanded than in the female, with the posterior edge nearly straight, and the last pair of uropoda appear a little more elongated and still more densely setous.

Occurrence. — This species would seem to be one of the most frequent Amphipoda of the Caspian Sea. It has been collected by Mr. Warpachowsky in no less than 14 different Stations, and in some of them in great abundance. Of these Stations, one (St. 2) is located off the Tschistyi-Bank, another (St. 12), in the inner part of the Bai Agrachansky, a 3rd (St. 40), off the promontory Brjanskaja, a 4th (St. 31), about midway between the peninsula Mangyschlak and the opposite western coast, a 5th (St. 61), far north, at some distance outside the Bai Bogatyi Kultuk, the remaining Stations (16, 17, 23, 24, 26, 27, 28, 51, 54), distributed over the tract north of the peninsula Mangyschlak. I have also had an opportunity of examining some specimens of this species preserved in the Museum of St. Petersburgh from earlier time, and collected by Baer and Göbel, partly at Baku, partly at Krasnowodsk.

Dr. Grimm has collected this species in several localities both of the southern and middle part of the Caspian Sea, from the shore down to a depth of 6 fathoms. A single specimen, that labelled *G. robustoides*, was found at the considerable depth of 108 fathoms.

Distribution. — To judge from the one of the specific names (*aralocaspius*) attributed by Dr. Grimm to this species, it would also seem to occur in the Aral Sea. I have not yet, however, had an opportunity of examining any specimens from that basin.

14. Gammarus crassus, Grimm MS.

(Pl. XIII).

Specific Characters. — Body rather short and stout, with broadly rounded back. Lateral lobes of cephalon somewhat projecting and obtusely rounded at the tip. Anterior pairs of coxal plates of moderate size, and fringed distally with scattered bristles: 1st pair but very slightly widening distally; 4th pair considerably expanded in their outer part, though of scarcely as broad as they are deep. The last 2 pairs of epimeral plates of metasome acutely produced at the lateral corners. Urosome smooth above, 1st segment with a dorsal fascicle of fine hairs, the 2 succeeding ones each with one or two spinules on either side of the dorsal face. Eyes of moderate size and oval reniform. Antennæ nearly equal-sized, and scarcely attaining to ⅓ of the length of the body; the superior ones with the 1st joint of the peduncle about the length of the other 2 combined, flagellum considerably longer than the peduncle, accessory appendage 4—5-articulate. Gnathopoda in female not very strong, in male much more powerfully developed, the posterior ones having the propodos very large. Anterior pairs of pereiopoda less robust than in *G. robustoides*; the 3 posterior pairs moderately elongated, with the basal joint rather expanded, that of last pair being very large, with the posterior edge distinctly serrate, and having the inferoposteal corner expanded to a broad, obtusely truncated lobe projecting far beyond the ischial joint. Last pair of uropoda nearly as in *G. robustoides*. Telson somewhat broader than it is long, each half with only 2 apical spines. Length of adult female 11 mm., of male 12 mm.

Remarks. — The present species, established by Dr. Grimm, is nearly allied to *G. robustoides*, differing, however, in the still stouter form of the body, the different armature of the urosome, the structure of the gnathopoda and the large size and peculiar form of the basal joint of the last pair of pereiopoda. It is also rather inferior in size.

Description of the female.

Adult, ovigerous specimens scarcely exceed 11 mm. in length.

The body (see fig. 1) is of a still shorter and stouter form than in *G. robustoides*, being rather tumid, with broadly vaulted back, and the species thus fully deserves the specific name proposed for it by Dr. Grimm.

The cephalon about equals in length the first 2 segments of the mesosome combined, and has the lateral lobes rather projecting and obtusely rounded at the tip, being defined behind by a rather deep emargination.

The anterior pairs of coxal plates are of moderate size, and have their distal edge slightly crenulated and fringed with scattered bristles. The 1st pair (see fig. 4) are very slightly expanded distally, and are obtusely rounded at the tip. The 2nd pair (see fig. 5) have the distal edge somewhat oblique, whereas the 3rd pair are more regularly oblong quadrangular in form. The 4th pair (see fig. 6) are rather broadly expanded in their outer part, though scarcely as broad as they are deep. Their distal edge is smooth in the middle and passes both into the anterior and posterior edges in an even curve.

The epimeral plates of the metasome are well developed, and the last 2 pairs acutely produced at the lateral corners. The urosome is of moderate size and does not exhibit any dorsal prominences. The 1st segment has dorsally a fascicle of delicate bristles but no spines. The last 2 segments, on the other hand, are armed on either side of the dorsal face with one or two small spinules.

The eyes are of the usual oval reniform shape, and have the visual elements well developed and the pigment dark.

The antennæ are comparatively short, though perhaps a little more elongated than in the preceding species, and are not very different in length. The superior ones (fig. 2) do not nearly attain to $^{1}/_{3}$ of the length of the body, and have the 1st joint of the peduncle rather large, being fully as long as the other 2 combined. The flagellum is very slender and considerably longer than the peduncle, being composed of about 16 articulations. The accessory appendage about equals in length the 2nd peduncular joint, and is composed of 4 articulations. The inferior antennæ are a little shorter than the superior and of the usual structure.

The gnathopoda (figs. 4, 5) are not nearly so strongly developed as in the female of *G. robustoides*, and are also less unequal. The propodos of the anterior ones (fig. 4) is oval in form, with the palm rather oblique and defined below by a very slight angle, carrying 2 strong spines. In the posterior pair (fig. 5) the propodos is somewhat more elongated and of an oval quadrangular form, the palm being nearly transverse.

The 2 anterior pairs of pereiopoda (fig. 6) are less strongly built than in *G. robustoides*, but are otherwise of a rather similar structure.

The 3 posterior pairs of pereiopoda (figs. 7—9) are moderately slender, and have their outer part edged with fascicles of spines and delicate bristles. In all the basal joint is rather expanded, though of very different size. In the antepenultimate pair (fig. 7) this joint is of a rounded quadrangular form, with the infero-posteal corner somewhat projecting; in the penultimate pair (fig. 8) it is somewhat larger, being strongly expanded posteriorly, with the hind edge boldly curved and distinctly serrate, each serration carrying a short bristle. The last pair (fig. 9) are highly distinguished by the very large size of the basal joint, which expands at the infero-posteal corner to a broad, obtusely truncated lobe, reaching nearly to the middle of the meral joint. The posterior edge of the joint is distinctly serrate throughout, and provided with a number of short bristles corresponding to the serrations. Of the outer joints in these legs, the carpal one is a little shorter than the propodal joint. The dactylus is in all very strong and curved.

The 2 anterior pairs of uropoda (fig. 10) have the rami nearly equal-sized and each tipped with several spines, one of which is longer than the others; the inner ramus has also one or two lateral spines, whereas the outer is without such spines.

The last pair of uropoda (fig. 11) nearly agree in their structure with those in *G. robustoides*.

The telson (fig. 12) is scarcely as long as it is broad, and has, on the tip of each half, 2 spines accompanied by 2 delicate hairs.

The adult male (fig. 13), as usual, attains a somewhat larger size than the female, the length of the body amounting to 12 mm. It is of a somewhat more slender and compressed form, and also easily recognizable by the strong development of the gnathopoda.

The latter (figs 14, 15) are very unequal, the posterior ones (fig. 15) being much more powerful than the anterior, with the propodos exceedingly large and tumid. In the anterior ones (fig. 14) the propodos is considerably narrower and obpyriform in shape, being scarcely more than half as large as that of the posterior. In both pairs the palm is rather oblique, and armed in the middle, outside, with a strong spine, in addition to those occurring on the inferior corner.

Occurrence. — Of this form numerous specimens were collected by Mr. Warpachowsky at Stat. 49, lying between the islands Kulaly and Morskoy. It also occurred, though more sparingly, in 5 other Stations (16, 21, 32, 54, 55) distributed over about the same tract of the North Caspian Sea.

In the collection of Dr. Grimm this species is rather abundantly represented, but only a single bottle, containing 3 very small and somewhat defective specimens taken from the considerable depth of 108 fathoms, bears the name of the species. The other specimens were collected in comparatively shallow water, partly at Baku, partly on the west coast of Sara, and partly near the mouth of the river Surgudschy.

15. Gammarus abbreviatus, G. O. Sars, n. sp.

(Pl. XIV).

Specific Characters. — Body short and robust, being rather tumid in the female. Lateral lobes of cephalon slightly prominent and broadly rounded at the tip. Anterior pairs of coxal plates of moderate size, and having their distal edge conspicuously crenulated and fringed with rather long bristles; 1st pair obliquely expanded distally; 4th pair very large, being fully as broad as they are deep. The last 2 pairs of epimeral plates of metasome acutely produced at the lateral corners. Urosome smooth above, with one or two very small spinules on the dorsal face of the 2 posterior segments. Eyes oval reniform. Antennae unusually short; the superior ones not attaining the length of the inferior, but having the 1st joint of the peduncle very large, flagellum about the length of the peduncle, accessory appendage 4—5-articulate. Gnathopoda in female not very strong, propodos in both pairs nearly of same form, with the palm rather oblique; those in male, as usual, much stronger and more unequal, the posterior ones being much the larger. Anterior pairs of pereiopoda very robust and densely setiferous; the 3 posterior pairs considerably more slender, basal joint of last pair regularly oval in form, with the infero-posteal corner slightly produced and narrowly rounded. Last pair of uropoda nearly as in the 2 preceding species. Telson fully as long as it is broad, each half with 3 apical spines. Length of adult female 12 mm., of male 13 mm.

Remarks. — This new species somewhat resembles *G. crassus* in the short and stout body, but is, on a closer examination, easily distinguishable by the unusually short superior antennæ, the different shape of the 1st and 4th pairs of coxal plates, and by the form of the basal joint of the last pair of pereiopoda. It also attains a somewhat larger size than that species.

Description of the female.

The length of a fully adult, ovigerous specimen measures 12 mm.

The form of the body (see fig. 1) is comparatively short and stout, with the anterior division rather tumid, and the back broadly rounded.

The cephalon does not attain the length of the first 2 segments of the mesosome combined, and has the lateral lobes slightly prominent and evenly rounded at the tip, being defined behind by a rather deep emargination.

The anterior pairs of coxal plates are of moderate size and have their distal edge conspicuously crenulated and fringed with rather long and slender bristles. The 1st pair (see fig. 5) are obliquely expanded in their outer part extending beneath the cephalon as far as the insertion of the inferior antennæ. The 4th pair (see fig. 7) are very large and expanded, being fully as broad as they are deep, and are vertically truncated below the posterior emargination.

The epimeral plates of the metasome are well developed, and of about same appearance as in the preceding species.

The urosome is rather short and quite smooth above, with only one or two very small spinules on the dorsal face of the 2 posterior segments.

The eyes are of the usual oval reniform shape.

The superior antennæ (fig. 3) are unusually short, not even attaining to $\frac{1}{4}$ of the length of the body, and have the 1st joint of the peduncle very large, exceeding in length the other 2 combined. The flagellum is scarcely longer than the peduncle, and is composed of about 9 articulations. The accessory appendage about equals in length the last 2 peduncular joints combined, and is composed of 4—5 articulations.

The inferior antennæ (fig. 4), unlike what is generally the case in this genus, are somewhat longer than the superior, and rather densely setous posteriorly. The flagellum about equals in length the last 2 peduncular joints combined, and is composed of about 7 articulations.

The gnathopoda (figs. 5, 6) are not very strong, but of the very same structure, though the posterior ones are a little larger. The propodos in both pairs is of a somewhat irregular oval form, with the palm rather oblique and defined below by a very slight angle armed with 2 spines.

The 2 anterior pairs of pereiopoda (fig. 7) are very strongly built and densely setiferous, with the meral joint large and expanded.

The 3 posterior pairs of pereiopoda (figs. 8—10) are moderately slender, and have their outer part edged with fascicles of slender bristles and scattered spines. The basal joint of the antepenultimate pair (fig. 8) exhibits the usual rounded quadrangular shape; that of the penultimate pair (fig. 9) is only expanded in its proximal part, and gradually tapers distally. The last pair (fig. 10) have the basal joint rather large and of a regularly oval form, with the infero-posteal corner projecting below as a narrowly rounded lappet. The posterior edge of this joint is, as in the 2 preceding pairs, minutely serrate and fringed with short bristles. Of the outer joints in these legs, the carpal one considerably exceeds the meral one in length, being fully as long as the propodal joint.

The 2 anterior pairs of uropoda (figs. 11, 12) have the rami rather narrow and provided at the tip with several spines; the outer ramus has also one, and the inner, 2 lateral spines.

The last pair of uropoda (fig. 13) do not differ much from those in the 2 preceding species, having the inner ramus very small, and the outer densely fringed with partly ciliated setæ.

The telson (fig. 19) is about as long as it is broad, with the lateral halves smooth except at the tip, which carries 3 slender spines.

The adult male (fig. 14) is a little larger than the female, the length of the body measuring about 13 mm.

It does not differ much in its external appearance from the female, except in being somewhat less tumid.

In the specimen examined the accessory appendage of the superior antennæ (fig. 15) was somewhat more fully developed and composed of 5 distinct articulations; otherwise the antennæ exhibited the very same structure as in the female.

The gnathopoda (figs. 16, 17) are, as usual, much more powerfully developed than in the latter, and of rather unequal size, the posterior ones (fig. 17) being much the stronger, with the propodos very large, and of an oblong oval form. In the anterior pair (fig. 16) the propodos is not nearly so large, but of a more obpyriform shape. In both pairs the palm is rather oblique, with the defining corner nearly obsolete, though armed with the usual spines. As in most other male Gammari, there is also found a spine about in the middle of the palm outside.

The last pair of uropoda (fig. 18) are somewhat larger than in the female, otherwise of a very similar structure.

Occurrence. — Of this form solitary specimens were collected by Mr.

Warpachowsky in 3 different Stations of the North Caspian Sea, the first (St. 54) located at some distance north of the islands Kulaly and Morskoj, the 2nd (St. 58) lying North of the Tschistyi-Bank, and the 3rd (St. 56) occurring about midway between the 2 former Stations.

In the collection of Dr. Grimm, only a single specimen of this form was found. It was, according to the label, taken off the west coast of Sara, among Zostera.

16. Gammarus obesus, G. O. Sars, n. sp.

(Pl. XV).

Specific Characters. — Body exceedingly short and stout, with broadly vaulted back. Lateral lobes of cephalon but slightly projecting, and evenly rounded at the tip. Anterior pairs of coxal plates very large and fringed on their distal edge with slender bristles; 1st pair rather widely expanded distally; 4th pair not nearly as broad as they are deep. Epimeral plates of metasome not very large, and scarcely produced at the lateral corners. Urosome short and stout, being quite smooth above. Eyes oval reniform. Superior antennae somewhat longer than the inferior, with the 1st joint of the peduncle rather large, flagellum a little longer than the peduncle, accessory appendage comparatively small. Gnathopoda in female not very strong, subequal, in male somewhat larger and more unequal. All pereiopoda very densely furnished with bristles, the 3 posterior pairs comparatively strongly built, with the carpal joint rather short; basal joint of antepenultimate and penultimate pairs of nearly same shape, with the infero-posteal corner not at all produced; that of last pair very large and expanded, forming at the infero-posteal corner a broadly rounded lobe. Last pair of uropoda unusually short and stout, inner ramus minute, outer ramus setiferous in its outer part only, terminal joint extremely small. Telson short and broad, each half with only a single apical spine. Length of adult female 8 mm., of male 9 mm.

Remarks. — The present new species is highly distinguished by its unusually short and compact form of body, differing in this point considerably from the Gammarid type, and approaching the species of the next genus. It is moreover easily recognized by the densely hirsute and rather strongly built pereiopoda, and by the poor development of the last pair of uropoda.

Description of the female.

The length of an apparently adult specimen measures about 8 mm.

The form of the body (see fig. 1) is extremely short and stout, indeed more so than in any other known Gammarus, with the back broadly vaulted and smooth throughout.

The cephalon about equals in length the first 2 segments of the mesosome combined, and has the lateral lobes but slightly prominent and evenly rounded at the tip.

The anterior pairs of coxal plates are unusually large and closely contiguous, being nearly twice as deep as the corresponding segments. They are fringed on the distal edge with rather long and slender setæ springing from small crenulations of the edge. The 1st pair (see fig. 4) are rather broadly expanded in their outer part, nearly concealing the buccal area at the sides. The 2 succeeding pairs (see fig. 5) are oblong quadrangular in form and of nearly same breadth throughout. The 4th pair (see fig. 6) are, as usual, larger than the preceding pairs, though not very much expanded in their outer part, being not nearly as broad as they are deep. Their posterior expansion is obliquely truncated and projects immediately below the emargination to an acute corner.

The epimeral plates of the metasome are not very large, and not at all produced at the lateral corners, the 1st pair being evenly rounded, the other two obtusangular.

The urosome is comparatively short and stout, and quite smooth above, with only a very small spinule on each side of the dorsal face of the last segment.

The superior antennæ (fig. 2) about equal in length $1/3$ of the body and have, as in the preceding species, the 1st joint of the peduncle rather large, exceeding in length the other 2 combined. The flagellum is somewhat longer than the peduncle, and composed of about 15 articulations. The accessory appendage is comparatively small and in the specimen examined consisted of only 2 articulations.

The inferior antennæ (fig. 3) are a little shorter than the superior, and have the flagellum about the length of the last 2 peduncular joints combined, and composed of 7 articulations.

The gnathopoda (figs. 4, 5) are of moderate size and almost exactly alike, though the posterior ones (fig. 5) are perhaps a little larger than the anterior (fig. 4). The carpus is in both pairs comparatively short and expanded below to a rounded setiferous lobe. The propodos is oval quadrangular in form, with the palm rather short, and defined below by an obtuse angle carrying a single spine.

The 2 anterior pairs of perciopoda (fig. 6) are moderately strong and densely setiferous, some of the setæ attached to the posterior edge of the meral joint being distinctly ciliated.

The 3 posterior pairs of perciopoda (figs. 7—9) are comparatively strongly built and very densely setiferous, with the carpal joint comparatively short, and the dactylus very strong. The basal joint of the antepenultimate pair (fig. 7) is of a somewhat unusual form, being not at all produced at the infero-posteal corner, but of nearly the same shape as that of the penultimate pair (fig. 8), though somewhat shorter. In both pairs this joint is densely fringed posteriorly with short setæ, and has anteriorly several fascicles of slender bristles. The last pair (fig. 9) have the basal joint very large and expanded, widening distally and forming at the infero-posteal corner a broadly rounded lobe. The posterior edge of the joint is throughout fringed with numerous rather slender and elongated setæ, and the anterior edge is also rather richly supplied with bristles arranged in several dense fascicles.

The 2 anterior pairs of uropoda (comp. figs. 16, 17) have the rami quite smooth, except at the tip, which is armed with a strong spine accompanied by 2 much smaller ones.

The last pair of uropoda (fig. 10) are unusually short and stout, but reaching little beyond the others. The basal part is rather thick and massive, being armed at the end below with several spines. The inner ramus exhibits the usual scale-like appearance, and carries a single small apical spine. The outer ramus is scarcely longer than the basal part, and provided in its outer part with a number of long ciliated setæ and with 2 spines on the outer edge. The terminal joint is so extremely minute as readily to escape attention.

The telson (fig. 11) is short and broad, being, as usual, divided by a deep and narrow cleft into two halves, each of which carries, at the narrowly truncated tip, a single spine accompanied by 2 small hairs.

The adult male (fig. 12) is somewhat larger than the female, measuring in length about 9 mm., and exhibits a similar short and compact form of the body.

In the specimen examined, the accessory appendage of the superior antennæ (fig. 13) was somewhat more fully developed than in the female, being composed of 4 articulations. It did not, however, much exceed the last peduncular joint in length.

The gnathopoda (figs. 14, 15) are, as usual, more strongly built than in the female, though the difference in this point is not as great as in most other species. The propodos in both pairs is oblong oval in form and in the posterior pair somewhat larger than in the anterior, the palm being in both somewhat more oblique than in the female.

The last pair of uropoda (fig. 18) are scarcely larger than in the latter but of a similar structure.

Occurrence. — Solitary specimens of this form were collected by Mr. Warpachowsky in 3 different Stations of the North Caspian Sea, the 1st

(St. 16) located off the island Swjatoj, the 2nd (St. 40) off the promontory Bramskaia, and the 3rd (St. 61) lying far north, outside the Bai Bogatui Kultuk.

Dr. Grimm's collection contains several specimens of this form, collected in quite shallow water, partly at Baku, partly at the mouth of the river Sargudschy.

Gen. 5. **Niphargoides**, G. O. Sars.

Syn.: *Niphargus*, Grimm (not Schödte).

Generic Characters. — Body smooth and of a very robust form, with the coxal plates not very large, and more or less densely setous on their distal edge. Cephalon comparatively small and without any rostrum, lateral lobes more or less projecting. Eyes distinct, though generally not very large. Antennæ extremely short and stout, equal-sized, and more or less densely setous, the superior ones provided with a distinct accessory appendage, and having their outer part, as a rule, extended laterally. Mandibular palps large, and generally densely setous. Oral parts otherwise normal. Gnathopoda strongly developed and of same appearance in the two sexes, though rather differing in shape in the different species, both pairs distinctly subcheliform. Pereiopoda rather strong and more or less densely clothed with bristles, basal joint of last pair much larger than that of the 2 preceding pairs. The 2 anterior pairs of uropoda comparatively strongly built, with the rami subequal; last pair not very large, with the inner ramus squamiform, the outer more or less densely setiferous, and having a very small terminal joint. Telson divided by a deep and narrow cleft into two halves spinous at the tip.

Remarks. — This genus is founded upon the form recorded by Dr. Grimm under the name of *Niphargus caspius*. In his treatise on some blind Amphipoda of the Caspian Sea, this author observes, that the above-named form might perhaps be more properly regarded as the type of a separate genus, differing, as it does, in some points rather conspicuously from the known species of the genus *Niphargus*, Schödte, though he believes that there is, in a genealogical sense, a near relationship between the two. In my opinion, the Caspian form ought, indeed, to be far removed from the genus *Niphargus*, to which in reality it does not exhibit any very close affinity, nearly all the appendages being very differently constructed. As I wish, however, to make as little change as possible with the names proposed by Dr. Grimm, I have substituted for the generic name *Niphargus* that of *Niphargoides*. In their outward appearance, the species of this new genus exhibit a much closer resemblance to another apparently widely remote genus, viz., that of *Pontoporeia*, and in some species, indeed, the similarity is truly perplexing; but the gnathopoda are very different. In many particulars the new genus would seem to approach nearer to the genus *Gammarus* than to that of *Niphargus*, and there are, as above stated, some species of the former genus, which form, as it were, a transition to the type revealed in the present genus.

In the collection of Mr. Warpachowsky, no less than 4 distinct species of this genus are to be found, one of which will be described below, the others in a subsequent article. A 5th species is also represented in the collection of Dr. Grimm.

17. Niphargoides caspius (Grimm).

(Pl. XVI).

Syn.: *Niphargus caspius*, Grimm.

Specific Characters. — Body somewhat elongated, but very tumid, with broadly vaulted back. Cephalon rather small, with the lateral lobes somewhat projecting and rounded at the tip. Anterior pairs of coxal plates but little deeper than the corresponding segments, and very densely clothed on their distal edge with slender bristles; 1st pair not expanded distally; 4th pair somewhat deeper than they are broad. Epimeral plates of metasome well developed, the last 2 pairs nearly rectangular, and having outside the lateral corners an oblique row of delicate bristles. Urosome smooth above. Eyes of moderate size and oval reniform, pigment dark. Antennæ rather densely setiferous, the superior ones about twice as long as the cephalon, and having the 1st joint of the peduncle very large and massive, 3rd joint extremely small, flagellum about the length of the last 2 peduncular joints combined, accessory appendage half the length of the flagellum and 3-articulate. Inferior antennæ with the flagellum extremely short, being scarcely longer than the last peduncular joint. Gnathopoda very unequal, the posterior ones being much larger than the anterior, propodos in both pairs conically tapering distally, with the palm very oblique, its defining angle being nearly obsolete. Pereiopoda densely setiferous, the 2 anterior pairs rather robust, the 3 posterior pairs more slender; basal joint of antepenultimate pair rather narrow and nearly of same shape as that of the penultimate pair; basal joint of last pair moderately expanded and of broadly oval form, being densely fringed with bristles both anteriorly and posteriorly. The 2 anterior pairs of uropoda with the rami rather stout and armed with unusually strong, blunt spines; last pair comparatively short, with the outer ramus rather broad and edged in its outer part with long ciliated setæ.

Telson with the lateral lobes sublinear and slightly diverging, each carrying at the tip 3 strong spines. Length of adult male 11 mm.

Remarks. — In the collection of Dr. Grimm, 2 nearly allied, but evidently distinct species are labelled as *Niphargus caspius*. For the larger species this specific name may be retained; for the other species, which is not contained in the collection of Mr. Warpachowsky, I propose the name of *Niphargoides Grimmi*. The species here described is easily distinguished by the densely hirsute anterior pairs of coxal plates and legs, the peculiar conically tapering shape of the propodos of both pairs of gnathopoda, as also by the form of the basal joint of the last pair of pereiopoda. Most of the specimens examined would seem to be of the male sex.

Description of the adult male. — The length of the largest specimen measures about 12 mm.

The form of the body (see fig. 1) is somewhat elongated, but rather robust, and not at all compressed, the breadth being fully as great as the height, including the coxal plates. The dorsal face is broadly vaulted and quite smooth throughout. In a dorsal view (fig. 2) the body exhibits a somewhat fusiform shape, with the greatest breadth occuring across the 4th segment of the mesosome, whence it gradually tapers both anteriorly and posteriorly. The metasome is well developed and about half the length of the anterior division of the body.

The cephalon is comparatively small, but little longer than the 1st segment of the mesosome, and does not exhibit any distinct rostral projection. The lateral lobes somewhat project between the insertions of the 2 pairs of antennæ, and are obtusely rounded at the tip. Behind they are defined by a very deep emargination, encircling the large and globular basal joint of the inferior antennæ.

The anterior pairs of coxal plates are not very large, being but little deeper than the corresponding segments, and are somewhat discontiguous in their outer part. They successively increase in size posteriorly, and have their distal edge very densely clothed with slender bristles in an almost brush-like manner. The 1st pair (see fig. 11) are about same breadth throughout, and obtusely rounded at the tip; the 2 succeeding pairs (see fig. 12) are a little narrowed distally, whereas the 4th pair (see fig. 14), as usual, are somewhat expanded in their outer part, forming, below the rather slight posterior emargination, an obtuse corner.

The 3 posterior pairs of coxal plates (see figs. 15—17) are unusually small and slightly bilobed.

The epimeral plates of the metasome are rather large, the 1st pair being, as usual, rounded, whereas the last 2 pairs are nearly rectangular, and having each, just above the lateral corner, outside, a somewhat oblique row of densely crowded delicate bristles (see fig. 1 a).

The urosome is of moderate size and quite smooth above, with only a very small spinule on either side of the dorsal face of the last segment.

The eyes are well developed, and of an oval reniform shape, with dark pigment.

The superior antennæ (fig. 3) are very short and stout, scarcely exceeding in length $\frac{1}{2}$ of the body, and about twice as long as the cephalon. They are rather richly supplied with bristles, and generally have their outer part extended laterally. The 1st joint of the peduncle is very large and massive, considerably exceeding in length the 2 outer joints combined; the last joint is extremely small, being scarcely longer than it is broad. The flagellum about equals in length the last 2 peduncular joints combined, and is composed of 7 rather short articulations. The accessory appendage is about half as long as the flagellum, and 3-articulate, the 1st articulation being about the length of the other 2 combined.

The inferior antennæ (fig. 4) are perhaps a little longer than the superior, and have the antepenultimate joint of the peduncle rather thick, forming posteriorly an angular, densely setous projection. The last joint of the peduncle is somewhat smaller than the penultimate, both being densely setiferous. The flagellum is extremely small, scarcely exceeding in length the last peduncular joint, and is composed of 5 articulations.

The buccal area (see fig. 1) considerably projects below, being only partly obtected at the sides by the 1st pair of coxal plates. The oral parts composing it are on the whole of normal structure resembling those in the genus *Gammarus*.

The anterior lip (fig. 5) exhibits the usual rounded form, and has the terminal edge somewhat narrowed and very slightly insinuated in the middle.

The posterior lip (fig. 7) has the inner lobes well defined, the outer ones projecting outside in a conical lappet.

The mandibles (fig. 6) are strongly built, and exhibit the usual armature of the masticatory part. The palp is very large, nearly twice as long as the mandible, and has the middle joint the largest, being edged inside with numerous long and slender bristles. The terminal joint is rather narrow and somewhat compressed, with the distal part of the inner edge densely setous, and having besides on the outer edge 3 fascicles of bristles.

The 1st pair of maxillæ (fig. 8) have the basal lobe well developed and of a triangular form, with about 8 ciliated setæ along the inner edge. The masticatory lobe and the palp are of the usual appearance.

The 2-nd pair of maxillæ (fig. 9) have the inner lobe smaller than the outer, both being provided at the tip with numerous curved bristles, and the inner lobe, in addition, with a series of setæ somewhat inside the edge.

The maxillipeds (fig. 10) are of moderate size, with the masticatory lobes a little larger than the basal ones, and armed on the inner edge with a row of strong spines, at the tip with several curved setæ. The palp is not very strong, with the last joint rather narrow, and the dactylus unguiform.

The gnathopoda (figs. 11, 12) are powerfully developed and very unequal in size, the posterior ones being much the larger. In structure they otherwise nearly agree with each other, both pairs having the basal joint rather strong and muscular and the carpus comparatively small, with a narrow setiferous lobe below. The propodos, especially in the posterior pair (fig. 12), is very large and elongated, gradually tapering distally, with the palm very oblique and scarcely defined from the hind margin by any distinct angle, though marked off from by 2 strong spines.

The 2 anterior pairs of pereiopoda (figs. 13, 14) are very strongly built and densely setiferous. The meral joint is considerably expanded, terminating in front in a projecting, densely setous corner, and also the carpal joint somewhat widens distally, whereas the propodal joint exhibits the usual narrow linear form.

The 3 posterior pairs of pereiopoda (figs. 15—17) are likewise rather strong and moderately elongated, being, like the anterior ones, densely covered with bristles both on the basal and terminal part. There also occur on the outer joints a number of strong spines, especially densely crowded at the end of the meral and carpal joints. The antepenultimate pair (fig. 15) are, as usual, somewhat shorter than the succeeding pairs, and have the basal joint not very large, oblong oval in form, and not at all produced at the infero-posteal corner. The basal joint of the penultimate pair (fig. 16) is somewhat more elongated, but otherwise of a similar appearance to that of the former pair. The last pair (fig. 17) have the basal joint much larger than in the preceding ones, and of a rather regular oval form, with the distal part of the anterior edge very densely clothed with bristles and projecting below as an obtuse corner. The posterior edge of the joint is quite evenly curved, and, as in the preceding pairs, densely fringed with rather long and slender bristles. Of the outer joints of these legs, the carpal one is fully as long as the propodal one, the dactylus not being very strong.

The 2 anterior pairs of uropoda (figs. 18, 19) are comparatively strongly built, with the rami subequal and rather stout, each having at the tip 4 unusually coarse and somewhat blunted spines, and a single lateral one.

The last pair of uropoda (fig. 20) are comparatively short and stout, but little projecting beyond the others. The basal part is rather thick and armed at the end below with a transverse row of strong spines. The inner ramus is very small and scale-like carrying, at the tip a single spine. The outer ramus is scarcely twice as long as the basal part and somewhat flattened, being fringed in its outer part with long, ciliated setæ, and having besides, about in the middle outside, 2 strong juxtaposed spines. The terminal joint is extremely small, nodiform.

The telson (fig. 21) consists of 2 slightly diverging lobes of nearly equal breadth throughout, each armed, on the obtusely truncated tip, with 3 strong spines increasing in length outwards.

Occurrence. — A few specimens of this form were collected by Mr. Warpachowsky at Stations 63 and 64, both lying in the eastern part of the North Caspian Sea.

Dr. Grimm has collected the species in 3 different Stations, one belonging to the southern part of the Caspian Sea, the other 2 to the middle part, the depth being from 35 to 40 fathoms.

EXPLANATION OF THE PLATES.

Pl. IX.

Gammarus Warpachowskyi, G. O. Sars.

Fig. 1. Adult, ovigerous female, viewed from left side.
» 2. Cephalon, from left side.
» 3. Superior antenna.
» 4. Inferior antenna.
» 5. Anterior gnathopod, with the corresponding coxal plate.
» 6. Posterior gnathopod, with coxal plate, branchial and incubatory lamellæ.
» 7. Second pereiopod, with coxal plate.
» 8. Antepenultimate pereiopod.
» 9. Penultimate pereiopod.
Fig. 10. Last pereiopod.
» 11. First uropod.
» 12. Second uropod.
» 13. Last uropod.
» 14. Telson.
» 15. Adult male, viewed from right side.
» 16. Inferior antenna of same.
» 17. Anterior gnathopod.
» 18. Posterior gnathopod.
» 19. Urosome, without the uropoda, viewed from right side.

Pl. X.

Gammarus minutus, G. O. Sars.

Fig. 1. Adult, ovigerous female, viewed from left side.
» 2. Superior antenna.
» 3. Inferior antenna.
» 4. Anterior gnathopod, with coxal plate.
» 5. Posterior gnathopod, with coxal plate, branchial and incubatory lamellæ.
» 6. Second pereiopod, with coxal plate.
» 7. Antepenultimate pereiopod.
Fig. 8. Penultimate pereiopod.
» 9. Last pereiopod.
» 10. Second uropod.
» 11. Last uropod.
» 12. Adult male, viewed from right side.
» 13. Anterior gnathopod of same, with coxal plate.
» 14. Posterior gnathopod, with coxal plate.
» 15. Last uropod.
» 16. Telson.

Gammarus macrurus, G. O. Sars.

Fig. 17. Adult, ovigerous female, viewed from left side.
» 18. Superior antenna.
» 19. Anterior gnathopod, with coxal plate.
» 20. Posterior gnathopod, with coxal plate, branchial and incubatory lamellæ.
» 21. Second pereiopod, with coxal plate.

Fig. 22. Antepenultimate pereiopod.
» 23. Penultimate pereiopod, without the propodal joint.
» 24. Last pereiopod, without the outer part.
» 25. First uropod.
» 26. Last uropod.
» 27. Telson.

Pl. XI.

Gammarus compressus, G. O. Sars.

Fig. 1. Adult ovigerous female, viewed from left side.
» 2. Superior antenna.
» 3. Anterior gnathopod, with coxal plate.
» 4. Posterior gnathopod, with coxal plate, branchial and incubatory lamellæ.
» 4a Outer part of the latter, more highly magnified.

Fig. 5. Second pereiopod, with coxal plate.
» 6. Antepenultimate pereiopod.
» 7. Penultimate pereiopod, without the outer joints.
» 8. Last pereiopod.
» 9. Last uropod.
» 10. Telson.

Gammarus similis, G. O. Sars.

Fig. 11. Adult, ovigerous female, viewed from right side.
» 12. Superior antenna.
» 13. Anterior gnathopod, with coxal plate.
» 14. Posterior gnathopod, with coxal plate, branchial and incubatory lamellæ.
» 15. Second pereiopod, with coxal plate.
» 16. Antepenultimate pereiopod.

Fig. 17. Basal part of penultimate pereiopod.
» 18. Last pereiopod.
» 19. Last uropod.
» 20. Telson.
» 21. Anterior gnathopod of male, with coxal plate.
» 22. Posterior gnathopod of same.

Pl. XII.

Gammarus robustoides, Grimm.

Fig. 1. Adult, ovigerous female, viewed from left side.
» 2. Superior antenna.
» 3. Anterior gnathopod, with coxal plate.
» 4. Posterior gnathopod, with coxal plate, branchial and incubatory lamellæ.
» 5. Second pereiopod, with coxal plate.
» 6. Antepenultimate pereiopod.
» 7. Penultimate pereiopod, without the outer joints.
» 8. Last pereiopod.
» 9. First uropod.

Fig. 10. Last uropod.
» 11. Telson.
» 12. Adult male, viewed from right side.
» 13. Accessory appendage of superior antenna.
» 14. Anterior gnathopod, with coxal plate.
» 15. Posterior gnathopod, with coxal plate, but without the branchial lamella.
» 16. Second uropod.
» 17. Urosome with telson, dorsal view.
» 18. Dorsal part of urosome, viewed from right side.

Pl. XIII.

Gammarus crassus, Grimm.

Fig. 1. Adult, ovigerous female, viewed from left side.
» 2. Superior antenna.
» 3. Dorsal part of urosome, viewed from left side.
» 4. Anterior gnathopod, with coxal plate.
» 5. Posterior gnathopod, with coxal plate, branchial and incubatory lamellæ.
» 6. Second pereiopod with coxal plate.

Fig. 7. Antepenultimate pereiopod.
» 8. Penultimate pereiopod.
» 9. Last pereiopod.
» 10. Second uropod.
» 11 Last uropod.
» 12. Telson.
» 13. Adult male, viewed from right side.
» 14. Anterior gnathopod.
» 15. Posterior gnathopod.

Pl. XIV.

Gammarus abbreviatus, G. O. Sars.

Fig. 1. Adult, ovigerous female, viewed from left side.
» 2. Dorsal face of urosome, viewed from left side.
» 3. Superior antenna.
» 4. Inferior antenna.
» 5. Anterior gnathopod, with coxal plate.
» 6. Posterior gnathopod, with coxal plate, branchial and incubatory lamellæ.
» 7. Second pereiopod, with coxal plate.
» 8. Antepenultimate pereiopod.
» 9. Basal part of penultimate pereiopod.
Fig. 10. Last pereiopod.
» 11. First uropod.
» 12. Second uropod.
» 13. Last uropod.
» 14. Adult male, viewed from right side.
» 15. Accessory appendage of superior antenna.
» 16. Anterior gnathopod, with coxal plate.
» 17. Posterior gnathopod.
» 18. Last uropod.
» 19. Telson.

Pl. XV.

Gammarus obesus, G. O. Sars.

Fig. 1. Adult female, viewed from left side.
» 2. Superior antenna.
» 3. Inferior antenna.
» 4. Anterior gnathopod, with coxal plate.
» 5. Posterior gnathopod, with coxal plate, branchial and incubatory lamellæ.
» 6. Second pereiopod, with coxal plate.
» 7. Antepenultimate pereiopod.
» 8. Penultimate pereiopod.
» 9. Last pereiopod.
Fig. 10. Last uropod.
» 11. Telson.
» 12. Adult male, viewed from right side.
» 13. Accessory appendage of superior antenna.
» 14. Anterior gnathopod, with coxal plate.
» 15. Posterior gnathopod.
» 16. First uropod.
» 17. Second uropod.
» 18. Last uropod.

Pl. XVI.

Niphargoides caspius, Grimm.

Fig. 1. Adult male, viewed from left side.
» 1 a Lateral corner of last epimeral plate of metasome.
» 2. Adult male, dorsal view.
» 3. Superior antenna.
» 4. Inferior antenna.
» 5. Anterior lip.
» 6. Right mandible, with palp.
» 7. Posterior lip.
» 8. First maxilla.
» 9. Second maxilla.
» 10. Maxillipeds, without the left palp.
Fig. 11. Anterior gnathopod, with coxal plate.
» 12. Posterior gnathopod, with coxal plate and branchial lamella.
» 13. First pereiopod.
» 14. Second pereiopod.
» 15. Antepenultimate pereiopod.
» 16. Penultimate pereiopod.
» 17. Last pereiopod.
» 18. First uropod.
» 19. Second uropod.
» 20. Last uropod.
» 21. Telson.

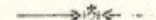

G.O.Sars Crustacea caspia.
Amphipoda. Pl.IX.

Gammarus Warpachowskyi, n. sp.

G.O.Sars Crustacea caspia.
Amphipoda. Pl X.

Figs 1–16 Gammarus minutus, n.sp. : deminutus
Figs. 17–27 Gammarus macrurus, n. sp.

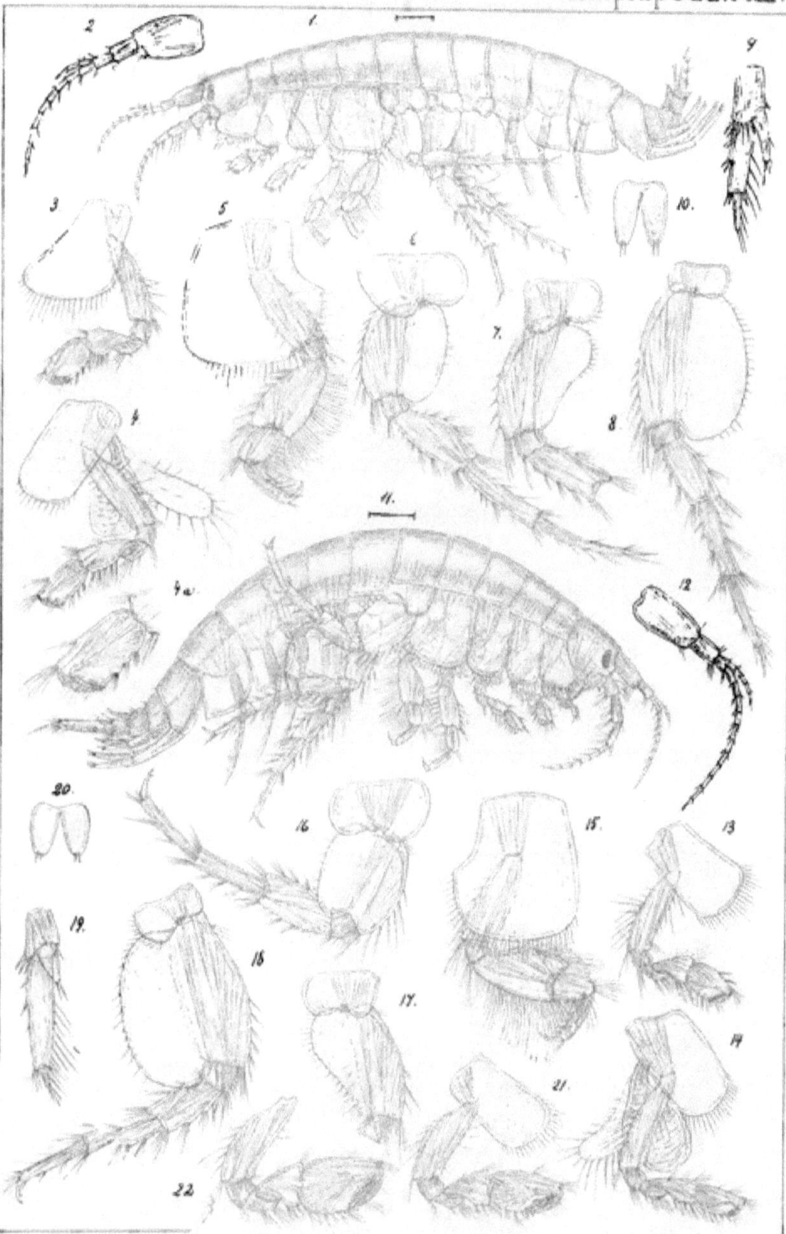

G.O.Sars Crustacea caspia.
Amphipoda. Pl. XI.

Figs. 1–10 Gammarus compressus, n. sp.
Figs. 11–20 Gammarus similis, n. sp.

G.O.Sars Crustacea caspia.
Amphipoda.Pl XII

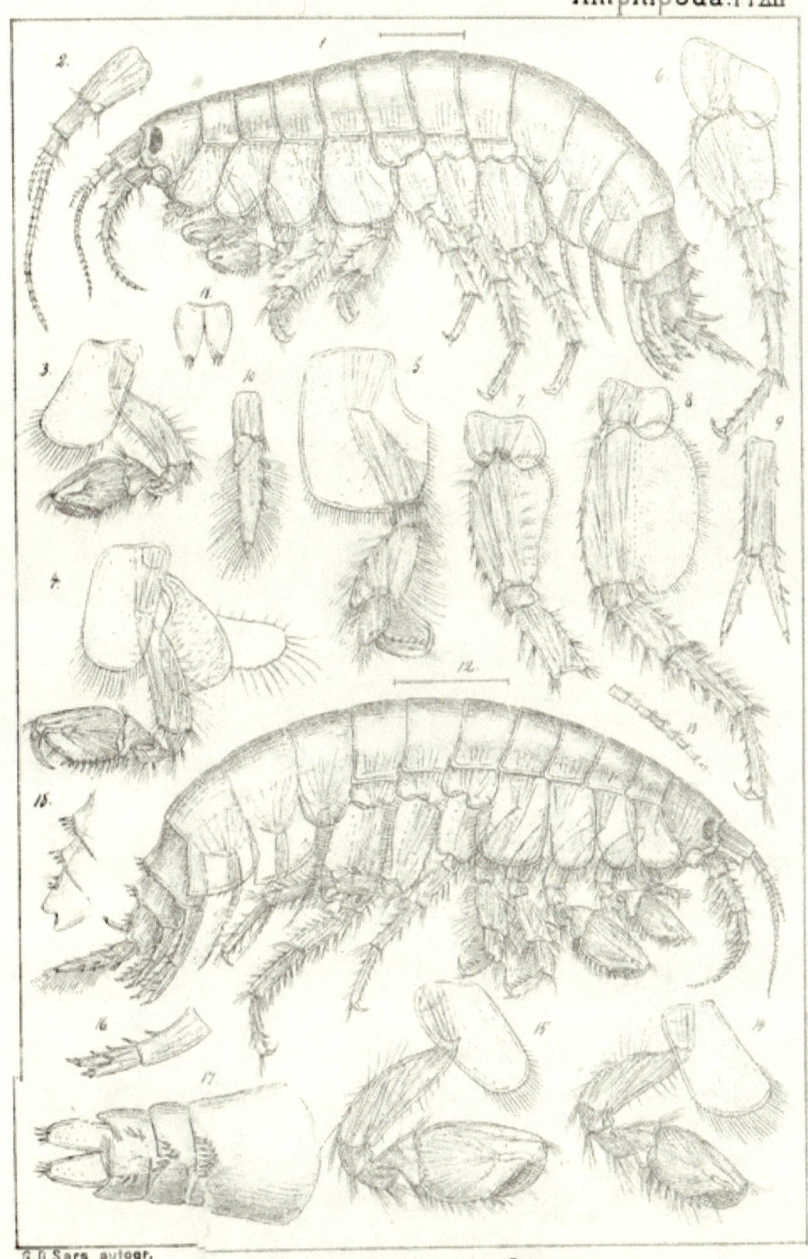

Gammarus robustoides Grimm.

G.O.Sars Crustacea caspia.
Amphipoda. Pl XIII

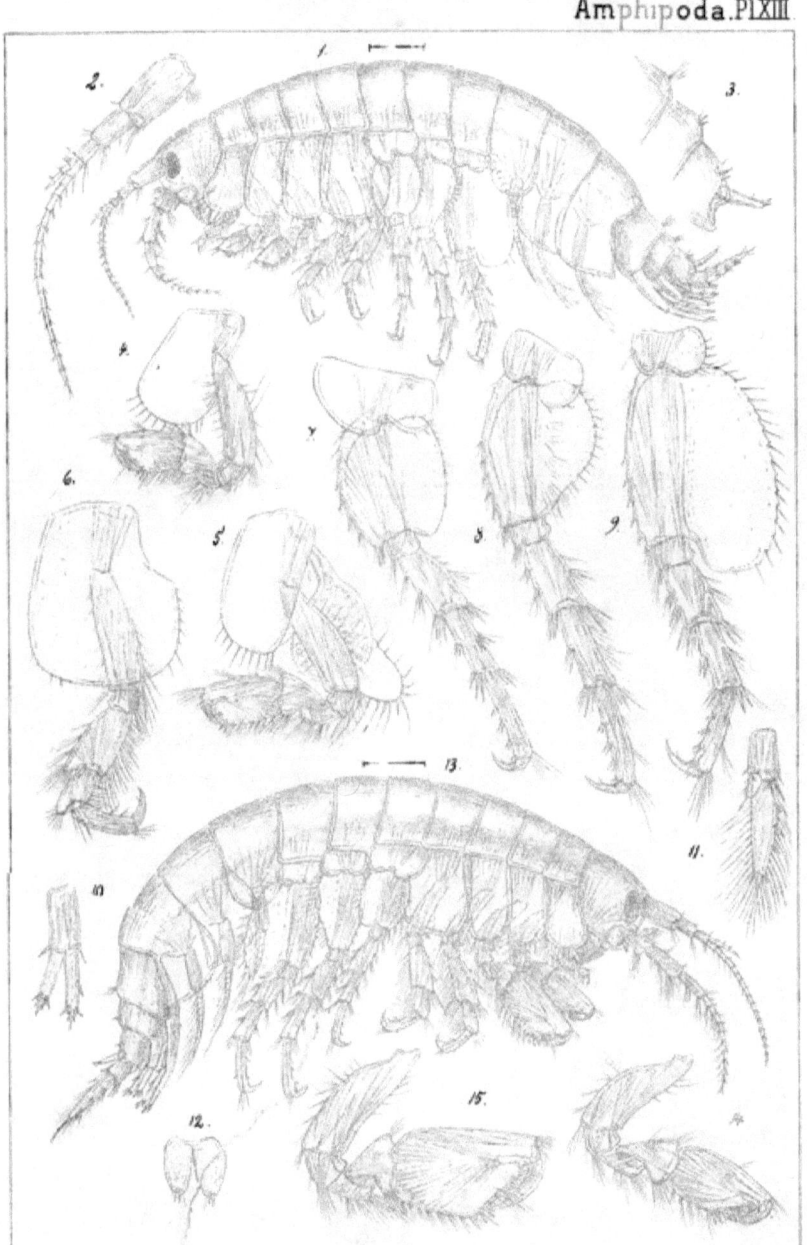

Gammarus crassus, Grimm.

G.O.Sars Crustacea caspia.
Amphipoda. Pl XIV.

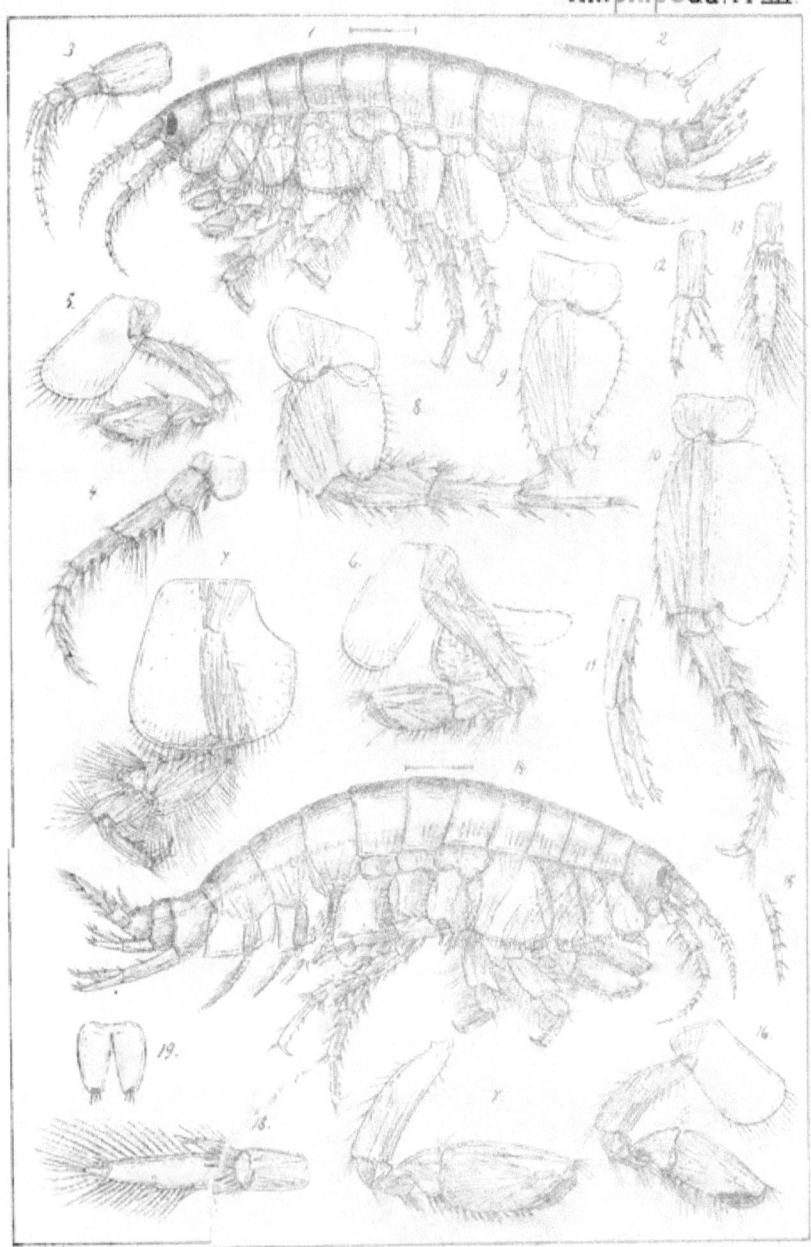

Gammarus abbrevialus, n.sp.

G.O.Sars Crustacea caspia.
Amphipoda. Pl. XV

Gammarus obesus, n.sp.

G.O.Sars Crustacea caspia.
Amphipoda. Pl. XVI.

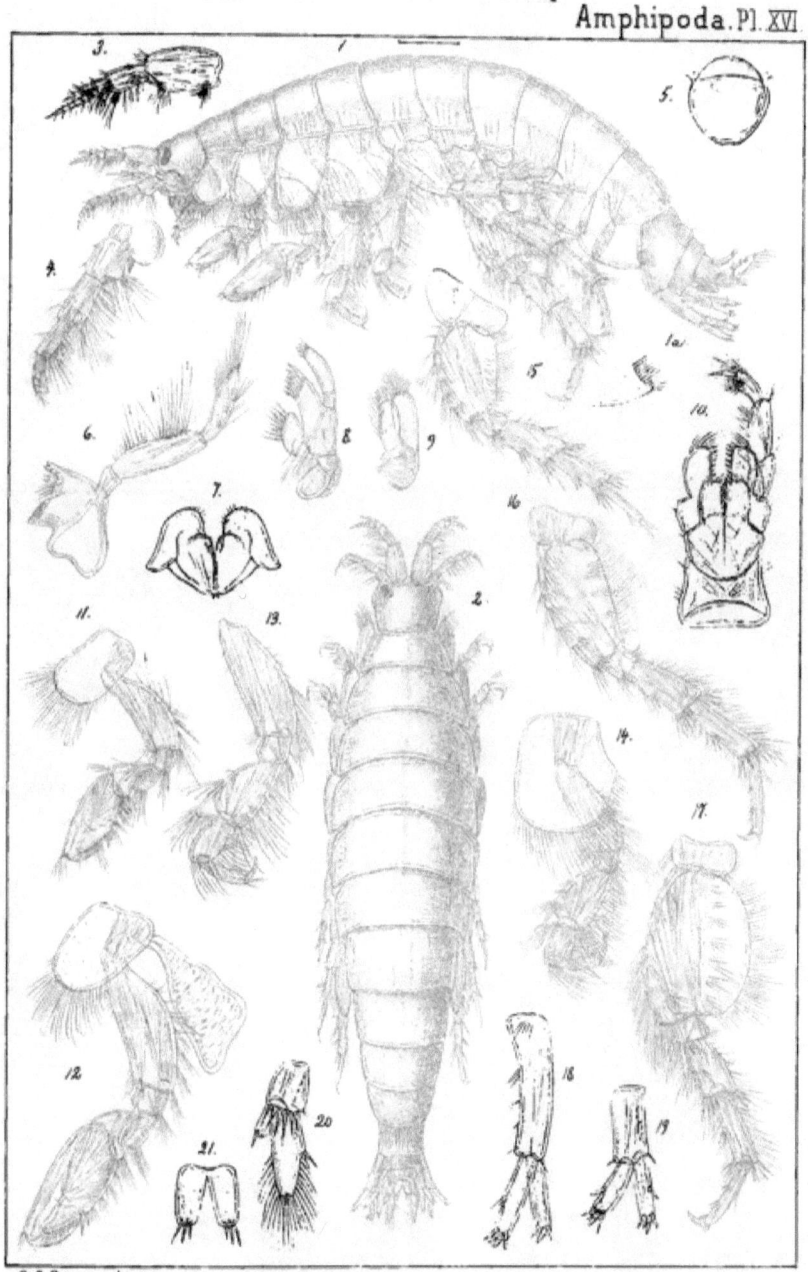

Niphargoides caspius, (Grimm.)

ОГЛАВЛЕНІЕ. — SOMMAIRE.

	Стр.
Извлеченія изъ протоколовъ засѣданій Академіи.	193
Н. Я. Сонинъ. О производныхъ функціяхъ высшихъ порядковъ.	321
Г. О. Сарсъ. Каспійскія ракообразныя. Матеріалы для изученія карцинологической фауны Каспійскаго моря (съ 8 таблицами рис.).	343
Г. Гильденъ. Преобразованіе періодическихъ агрегатовъ.	379
Кн. Б. Голицынъ. О свободной энергіи.	387

	Pag.
Extraits de procès verbaux des séances de l'Académie.	193
N. Sonin. Sur les dérivées d'ordre supérieur.	321
G. O Sars. Crustacea Caspia. Contributions to the knowledge of the Carcinological Fauna of the Caspian Sea (with 8 autographic plates).	343
H. Gyldén. Zur Transformation der periodischen Aggregate.	379
Pr. B. Galitzine. Sur l'énergie libre.	387

Напечатано по распоряженію Императорской Академіи Наукъ.
Декабрь 1894 г. Непремѣнный секретарь, Академикъ Н. Дубровинъ.

ТИПОГРАФІЯ ИМПЕРАТОРСКОЙ АКАДЕМІИ НАУКЪ.
Вас. Остр., 9 линія, № 12.

ИЗВѢСТІЯ
ИМПЕРАТОРСКОЙ АКАДЕМІИ НАУКЪ.

ТОМЪ III. № 3.

1895. ОКТЯБРЬ.

BULLETIN
DE
L'ACADÉMIE IMPÉRIALE DES SCIENCES
DE
ST.-PÉTERSBOURG.

Vᵉ SÉRIE. TOME III. № 3.

1895. OCTOBRE.

С.-ПЕТЕРБУРГЪ. 1895. ST.-PÉTERSBOURG.

Продается у коммиссіонеровъ Императорской Академіи Наукъ:	Commissionnaires de l'Académie Impériale des Sciences:
И. И. Глазунова, М. Эггерса и Комп. и К. Л. Риккера въ С.-Петербургѣ,	J. Glasounof, M. Eggers & Cie. et C. Ricker à St.-Pétersbourg,
Н. П. Карбасникова въ С.-Петербургѣ, Москвѣ и Варшавѣ,	N. Karbasnikof à St.-Pétersbourg, Moscou et Varsovie,
Н. Киммеля въ Ригѣ,	N. Kymmel à Riga,
Фосса (Г. Гессель) въ Лейпцигѣ.	Voss' Sortiment (G. Haessel) à Leipzig.

Цѣна: 1 р. — Prix: 2 Mk. 50 Pf.

Crustacea caspia.
Contributions to the knowledge of the Carcinological Fauna of the Caspian Sea.

By **G. O. Sars**,

Professor of Zoology at the University of Christiania, Norway.

Part III.

AMPHIPODA.

Third Article.

Gammaridæ (concluded). **Corophiidæ**.

With 8 autographic plates.

(Présenté le 19 avril 1895).

18. **Niphargoides corpulentus**, G. O. Sars, n. sp.

(Pl. XVII, figs. 1—19).

Specific Characters. — ♀. Body very robust and tumid, with broadly vaulted back. Cephalon exceeding in length the 1st segment of mesosome, lateral lobes obtusely rounded. Anterior pairs of coxal plates somewhat deeper than the corresponding segments, and fringed on the distal edge with moderately long bristles; 1st pair scarcely expanded distally; 4th pair about as broad as they are deep. Last pair of epimeral plates of metasome slightly produced at the lateral corners, and having outside the latter an oblique row of bristles. Segments of urosome slightly raised dorsally, the last 2 with a pair of small, subdorsal spinules. Eyes well developed, though not very large, oval reniform, pigment dark. Antennæ short and stout, the superior ones about twice the length of the cephalon, with the flagellum fully as long as the last 2 peduncular joints combined, accessory appendage half the length of the flagellum and 4-articulate. Inferior antennæ about the length of the superior, flagellum longer than the last peduncular joint. Gnathopoda moderately strong and somewhat unequal, the posterior ones being the larger, propodos in both pairs oblong oval, not tapering distally, palm well defined and shorter than the hind margin. Pereiopoda densely setiferous, the 2 anterior pairs very robust, with the meral and carpal joints lamellarly expanded; the 3 posterior pairs more slender, basal joint of last pair very large and expanded, with the posterior edge strongly arcuate and fringed with long setæ. The 2 anterior pairs of uropoda with the rami subequal and armed with spines of the usual shape. Last pair of uropoda comparatively short, outer ramus fringed with long ciliated setæ, inner ramus small, scale-like. Telson with the lateral lobes but slightly divergent and obtusely truncated at the tip, each with a row of 5 slender apical spines. Length of adult male 14 mm.

Remarks. — The present species is allied to *N. caspius*, but easily distinguishable by the more robust form of the body, the less densely hirsute coxal plates, the rather different shape of the propodos of the gnathopoda, and finally by the greatly expanded basal joint of the last pair of pereiopoda.

Description of the adult male. — The length of the largest specimen measures 14 mm., and this form accordingly grows to a considerably larger size than *N. caspius*.

The form of the body (see fig. 1) is very robust and tumid, with the dorsal face broadly vaulted, and the species thus fully deserves its specific name *corpulentus*.

The cephalon is of a shape similar to that in *N. caspius*, though exceeding somewhat in length the 1st segment of the mesosome. The frontal edge is but very slightly produced between the bases of the superior antennæ. The lateral lobes are somewhat projecting and obtusely rounded at the tip.

The anterior pairs of coxal plates are comparatively larger than in *N. caspius*, being considerably deeper than the corresponding segments, and are fringed on their distal edge with a regular row of bristles, which, however, are not nearly so much elongated and so densely crowded as in the said species. The plates successively increase in size posteriorly, the 1st pair (see fig. 4) being the smallest and of a regular oblong quadrangular form, with the outer part not expanded. The 2nd pair (see fig. 1) are somewhat narrowed distally, whereas the 3rd pair are almost of equal breadth throughout. The 4th pair are rather large, about as broad as they are deep, and are considerably expanded in their outer part, forming below the posterior emargination a rather projecting corner.

The 3 posterior pairs of coxal plates are, as in *N. caspius*, small and slightly bilobed, successively diminishing in size posteriorly.

The epimeral plates of the metasome are well developed, and of a shape similar to that in *N. caspius*. As in that species, the last pair (fig. 10) are provided with an oblique row of densely crowded bristles outside the lateral corners, which latter appear slightly produced.

The urosome is rather stout, and has the segments slightly raised dorsally, without, however, forming any distinct projections. They have a few small hairs on the upper face, and the last 2 segments besides 2 very small subdorsal spinules.

The eyes are well developed, though not very large, and of an oval reniform shape. The pigment in 2 of the 3 specimens examined was of the usual dark hue, in the 3rd, belonging to the collection of Dr. Grimm, it was almost quite absent, probably owing to the action of the spirit.

The superior antennæ (fig. 2) are short and stout, about twice as long as the cephalon, and have the 1st peduncular joint very large and massive, whereas the 3rd joint is rather small, scarcely exceeding half the length of the 2nd. The flagellum is a little longer than the last 2 peduncular joints combined and composed of 8 articulations. The accessory appendage is half as long as the flagellum, and 4-articulate.

The inferior antennæ (fig. 3) scarcely exceed in length the superior, and are less densely setous than in *N. caspius*, otherwise of a quite similar structure. The flagellum is, however, less rudimentary, exceeding in length the last peduncular joint, and is composed of 6 articulations.

The oral parts exactly agree in their structure with those in the type species.

The gnathopoda (figs. 4 and 5), on the other hand, are rather different, being on the whole less powerfully developed than in that species. They are somewhat unequal in size, the posterior ones (fig. 5) being, as usual, the larger, and are clothed with scattered fascicles of slender bristles. The propodos in both pairs is of a rather regular oblong quadrangular shape, being not, as in *N. caspius*, conically tapered distally. The palm is much less oblique, and is defined below by a distinct angle carrying 2 or 3 spines, the outmost of which is rather strong. The hind margin is considerably longer than the palm, and provided in the posterior pair with 3 fascicles of short spiniform bristles.

The 2 anterior pairs of pereiopoda (fig. 6) are very largely developed and densely setiferous. The meral joint is large and gradually expanded distally, projecting at the end anteriorly to an obtuse, densely setiferous projection. The carpal joint is very broad and lamellarly expanded, carrying on the posterior edge a regular series of strong curved setæ, and at the anterior corner a dense brush of slender bristles. The propodal joint, as in *N. caspius*, is rather narrow and setous only at the tip. The dactylus is about half the length of that joint.

The 3 posterior pairs of pereiopoda (figs. 7—9) are more slender than the anterior, and are, like the latter, rather densely setous, having, in addition to the setæ, at the end of the joints slender spines. The antepenultimate pair (fig. 7) are, as usual, considerably shorter than the other 2, which are about of equal length. The basal joint of the former is rather broad and obliquely oval in shape, with the anterior edge considerably curved. The basal joint of the penultimate pair (fig. 8) is considerably narrower and more elongated, with the posterior edge slightly sinuate in the middle and fringed with slender setæ. The last pair (fig. 9) are distinguished by the large size of the basal joint, which is greatly expanded and of a broad cordiform shape, with the posterior edge strongly curved below the middle, and fringed with long setæ springing off from small serrations of the edge. The outer joints of these legs nearly agree in their longitudinal relation with those in *N. caspius*.

The 2 anterior pairs of uropoda (fig. 11) are rather stout, but otherwise of quite normal structure, with the rami subequal and armed with scattered spines of the usual kind.

The last pair of uropoda (fig. 12) are comparatively short, and resemble in structure those in the type species. The basal part is short and thick, and is armed at the end below with a transverse row of 7 slender spines. The outer ramus is about twice as long as the basal part, and densely fringed in its outer part with ciliated setæ, having besides on the outer edge 2 strong spines. The terminal joint of this ramus is extremely small, nodiform. The inner ramus exhibits the usual scale-like shape, and scarcely exceeds in length the basal part. It is armed at the tip with 2 strong spines, and has inside 3 small bristles.

The telson (fig. 13) is, as in the other species, cleft to the base, being accordingly divided into 2 halves, which are somewhat longer and less diverging than in *N. caspius*. Each lobe carries at the obtusely truncated tip a transverse row of 5 slender spines increasing in length outwards, but is otherwise quite unarmed.

Occurrence. — Of this species 2 specimens were collected by Mr. Warpachowsky, the one at Stat. 2, in the western part of the North Caspian Sea, south of the T-chistyi Bank, the other at Stat. 59, farther north, at some distance from the mouth of the Wolga. Both specimens were of the male sex. A third male specimen has been collected, according to the label, by Dr. Baer, but without statement of locality.

19. Niphargoides compactus, G. O. Sars, n. sp.

(Pl. XVII. figs. 14—19).

Specific Characters. — ♂. Body extremely robust and compact, having the last 2 segments of mesosome and those of metasome each provided with a well-marked transverse sulcus dorsally. Cephalon comparatively small, with the lateral lobes evenly rounded. Anterior pairs of coxal plates rather large, fully twice as deep as the corresponding segments, and fringed distally with moderately long bristles; 1st pair considerably expanded in their outer part; 4th pair very large, deeper than they are broad. Last pair of epimeral plates of metasome about as in the preceding species. Urosome of moderate size; 2nd segment with a single small spinule dorsally; 3rd segment with 2 spinules on each side of the dorsal face. Eyes well developed, oval reniform. Antennæ short, subequal in length, the superior ones with the 2nd joint of the peduncle rather elongated, flagellum extremely small, accessory appendage 4-articulate. Inferior antennæ with the flagellum very small, not even attaining the length of the last peduncular joint. Gnathopoda very powerfully developed and rather unequal in size, propodos in both pairs large and broad at the base, obpyriform, with the palm very oblique and much longer than the hind margin, being defined below, in the posterior pair, by a distinct projecting angle armed with a strong spine. Pereiopoda nearly of the same structure as in *N. corpulentus*. Last pair of uropoda comparatively more fully developed than in the 2 preceding species, outer ramus sublamellar and densely fringed with ciliated setæ, inner ramus scale-like, having inside a row of ciliated bristles, and terminating with 2 small spines. Telson with each of the lateral halves armed at the obtusely truncated tip with 4 spines. Length of adult male 17 mm.

Remarks. — This new species is at once distinguished by its unusually stout and compact body, and by the distinct transverse sulci crossing the dorsal face of some of the segments. Moreover the structure of the antennæ and especially that of the gnathopoda may serve to easily recognize the species. I have only seen a single specimen, and for this reason have not been able to examine the oral parts. But there cannot be any doubt that it is congeneric with the 2 preceding species.

Description of the male. — The length of the specimen examined measures 17 mm., and this form accordingly grows to a much larger size than any of the other known species of the present genus.

The form of the body (see fig. 14) is extremely robust and compact, more so indeed than in any of the other species. The back is very broad and has across each of the 2 posterior segments of the mesosome and those of the metasome a very conspicuous transverse depression or sulcus.

The cephalon is comparatively small, scarcely longer than the 1st segment of the mesosome, and its lateral parts are partly concealed by the largely developed 1st pair of coxal plates. The lateral lobes are somewhat projecting and quite evenly rounded at the tip.

The anterior pairs of coxal plates are comparatively large, being fully twice as deep as the corresponding segments, and are fringed on their distal edges with a regular row of moderately long bristles. The 1st pair are, unlike what is the case in the 2 preceding species, considerably expanded in their outer part, being accordingly much broader than the succeeding pair. The latter are, like the 3rd pair, obliquely rounded at the tip, both pairs being almost exactly of the same shape, though somewhat differing in size. The 4th pair are very large and expanded, being somewhat deeper than they are broad, and exhibit the usual irregular, angular shape, with a distinctly projecting corner below the posterior emargination.

The 3 posterior pairs of coxal plates are comparatively small, though a little larger than in the 2 preceding species.

The epimeral plates of the metasome exhibit almost exactly the same shape as in *N. corpulentus*, and the last pair have a similar oblique row of bristles outside the lateral corners as found in the 2 preceding species.

The urosome is somewhat less robust than in *N. corpulentus*, but otherwise exhibits a very similar appearance. As in that species, there is a small dorsal spinule on the 2nd segment, and on the last segment (see fig. 19) 2 similar spinules are found on each side of the dorsal face.

The eyes are well developed and of a form and size similar to those in *N. corpulentus*.

The antennæ are short and subequal in length, being about twice as long as the cephalon. They are rather richly supplied with bristles, generally arranged in distinct fascicles, especially along the outer edge. The superior ones (fig. 15) have the 1st joint of the peduncle large and somewhat flattened, the second much narrower and rather elongated, whereas the 3rd joint is extremely small, scarcely exceeding $^1/_4$ of the 2nd. The flagellum is likewise unusually small, not even attaining half the length of the 2 last peduncular joints combined, and is composed of 9 articulations. The accessory appendage is about half as long as the flagellum, and 4-articulate. The inferior antennæ (fig. 16) have the 2 outer joints of the peduncle comparatively more slender than in the 2 preceding species, and densely clothed posteriorly with slender bristles. The flagellum is extremely small, being much shorter than the last peduncular joint, and is composed of 6 articulations.

The gnathopoda (figs. 17 and 18) are very powerfully developed and rather unequal in size, the posterior ones being much the stronger. The propodos in both pairs, but especially in the posterior one, is very large and greatly tumefied at the base, nearly obpyriform in shape, with the palm very oblique and much longer than the hind margin. The defining angle is on the posterior pair (fig. 18) greatly projecting and, as in the anterior pair, armed with a strong spine, which is accompanied by 2 smaller ones. The carpus is short and broad, being produced below to a narrow setiferous lobe.

The pereiopoda (see fig. 14) are on the whole very similar to those in *N. corpulentus*, and, as in that species, the basal joint of the last pair is very large and laminar, being densely fringed with bristles.

The 2 anterior pairs of uropoda are of the usual structure.

The last pair of uropoda (see fig. 19) appear somewhat more fully developed than in the 2 preceding species, and have the outer ramus rather broad, sublamellar, and densely fringed with ciliated setæ. As in the preceding species, there are besides on the outer edge of this ramus 2 ledges, to each of which are secured 2 spines. The terminal joint is so very small as easily to escape attention. The inner ramus exhibits the usual scale-like appearance and has inside a row of 7 short, ciliated setæ, at the tip 2 small spines.

The telson (ibid.) resembles that in *N. corpulentus*, except that each of the lateral halves has only 4 apical spines.

Occurrence. — The above described specimen was taken by Mr. Warpachowsky last summer in the eastern part of the North Caspian Sea, at Stat. 65.

20. Niphargoides quadrimanus, G. O. Sars. n. sp.

(Pl. XV, figs. 1—13).

Specific Characters. — Body less robust than in the 3 preceding species, and not nearly so tumid, back quite smooth throughout. Cephalon rather small, with the lateral lobes broadly rounded. Anterior pairs of coxal plates of moderate size, and fringed distally with a regular row of bristles; 1st pair scarcely expanded distally; 4th pair about as broad as they are deep. Last pair of epimeral plates of metasome nearly rectangular, and without any row of bristles outside the lateral corners. Eyes comparatively small, oval reniform. Antennæ comparatively more elongated than in the 3 preceding species and subequal in length, the superior ones with the 1st joint of the peduncle very large, fully twice as long as the other 2 combined, flagellum exceeding half the length of the peduncle, accessory appendage 6-articulate. Inferior antennæ rather strongly built, with the antepenultimate and penultimate joints of the peduncle expanded posteriorly to setiferous lobes, the outer 2 peduncular joints being moreover armed with spines arranged in oblique rows, flagellum exceeding half the length of the peduncle. Gnathopoda of exactly same appearance in the 2 sexes, being rather powerful and somewhat unequal in size, propodos of the anterior ones oval quadrangular, that of the posterior considerably larger and more regularly quadrate in outline, palm in both pairs nearly transverse, defining angle armed with 3 spines, the outmost of which is particularly strong. Anterior pairs of pereiopoda less robust than in the 3 preceding species, carpal joint scarcely expanded, propodal joint armed with a double row of slender spines. The 3 posterior pairs of pereiopoda rather much elongated and densely supplied with bristles as also with fascicles of slender spines; basal joint of last pair very much expanded, with the posterior edge somewhat irregularly curved and fringed with short bristles. The 2 anterior pairs of uropoda rather robust and armed with strong spines. Last pair of uropoda reaching considerably beyond the others, outer ramus more than twice as long as the basal part and edged with scattered non-ciliated bristles, terminal joint well defined; inner ramus small, scale-like. Telson small, with the lateral lobes strongly diverging, each with a single apical spinule. Length of adult female 10 mm.

Remarks. — The present form is chiefly characterised by the shape of the propodos of the posterior gnathopoda, which is more pronouncedly quadrate than in any of the other known species; hence the specific name. From the 3 preceding species it is moreover easily distinguished by its less robust body and by the structure of the antennæ and caudal appendages. In outer appearance this and the following species bear a strange resemblance to the species of the genus *Pontoporeia*.

Description of the female. — The length of fully adult, ovigerous specimens is about 10 mm.

The body (see fig. 1) is on the whole considerably more slender than in the 3 preceding species, and also much less tumid, with the back evenly rounded and quite smooth throughout, without any trace of the transverse depressions found in *N. compactus*.

The cephalon is comparatively small, though somewhat exceeding in length the 1st segment of the mesosome. The frontal edge is somewhat produced between the bases of the superior antennæ, without, however, forming any distinct rostral projection. The lateral lobes are rather prominent and broadly rounded at the tip; behind them there is a rather deep emargination encircling the large and swollen basal joint of the inferior antennæ.

The anterior pairs of coxal plates are of moderate size, being somewhat deeper than the corresponding segments, and are fringed distally with slender setæ, which become rather short on the 4th pair. The 1st pair (see fig. 4) are of about the same breadth throughout, and have the distal edge somewhat oblique. The 2 succeeding pairs are regularly oblong quadrangular in shape. The 4th pair (see fig. 6) are, as usual, the largest, being about as broad as they are deep, and exhibiting a distinctly projecting corner just below the posterior emargination. The 3 posterior pairs are small and bilobed.

The epimeral plates of the metasome are well developed and quite smooth. The last pair are nearly rectangular, and do not exhibit any trace of the oblique row of bristles found in the 3 preceding species outside the lateral corners.

The urosome is of moderate size and perfectly smooth above.

The eyes are distinct, though not very large, and of an oval reniform shape, with dark pigment.

The superior antennæ (fig. 2) are considerably more elongated than in the 3 preceding species, being about 3 times as long as the cephalon. The 1st joint of the peduncle is very large, fully twice as long as the other 2 combined, and is densely setous on the outer edge. The 3rd joint is about half as long as the 2nd, both being densely setous outside. The flagellum considerably exceeds half the length of the peduncle, and is composed of about 11 articulations. The accessory appendage is half as long as the flagellum, and 6-articulate.

The inferior antennæ (fig. 3) are about equal in length to the superior, and are rather strongly built, being generally bent in a genicular manner. The basal joint is very large and globular. The antepenultimate and penultimate joints of the peduncle are both expanded posteriorly to short setiferous lobes, that of the penultimate joint having, moreover, outside 2 oblique rows of short spines. The last peduncular joint is simple cylindric and nearly as long as the penultimate one. It has posteriorly several fascicles of slender bristles and outside 4 oblique rows of small spines. The flagellum is fully as long as the 2 outer joints of the peduncle combined, and is composed of 10 articulations.

The gnathopoda (figs. 4, 5) are rather powerful and somewhat unequal in size, the posterior ones (fig. 5) being, as usual, the larger. The propodos of the anterior gnathopoda (fig. 4) is quadrangular in shape, that of the posterior ones (fig. 5) considerably broader and more pronouncedly quadrate in outline. In both pairs the palm is nearly transverse and defined below by a distinct angle, to which are secured 3 spines, the outmost of which is particularly strong. The hind margin is somewhat longer than the palm, and exhibits in its outer part 3 or 4 fascicles of short bristles.

The 2 anterior pairs of pereiopoda (fig. 6) are moderately strong, with the meral joint rather large and densely setiferous on the posterior edge. The carpal joint is, on the other hand, but very little expanded, and is provided posteriorly, in addition to the setæ, with 3 strong spines. The propodal joint is, as usual, narrow linear, and is armed in its outer part posteriorly with a double row of slender spines.

The 3 posterior pairs of pereiopoda (figs. 7—9) are rather elongated and generally strongly reflexed. They have the outer part densely setiferous and besides provided with fascicles of slender spines. The antepenultimate pair (fig. 7) are, as usual, somewhat shorter than the other 2, and have the basal joint regularly oval in form, with from 4 to 5 fascicles of slender bristles anteriorly. The meral joint of this pair is rather broad, its posterior edge bulging considerably in the middle. In the penultimate pair (fig. 8) the basal joint is comparatively narrower and more elongated, with the posterior edge slightly sinuated below the middle. The last pair (fig. 9) are distinguished by the large size of the basal joint, which forms posteriorly a very broad lamellar expansion, the edges of which are somewhat irregularly curved and throughout fringed with short bristles. Anteriorly this joint terminates in an obtuse corner very densely clothed with slender bristles. The outer joints of these legs exhibit a similar longitudinal relation as in the 3 preceding species.

The 2 anterior pairs of uropoda (figs. 10—11) are rather strongly built, with the rami subequal and armed with 5 strong apical spines and a single lateral one.

The last pair of uropoda (fig. 12) are considerably more elongated than in the 3 preceding species, projecting far beyond the other pairs. The basal joint is rather short and armed at the end below with a transverse row of 5 not very elongated spines. The outer ramus is fully twice as long as the basal part and rather narrow, with only scattered simple bristles and 2 fascicles of spines on the outer edge. The terminal joint of this ramus is well defined and about ⅟, as long as the proximal one, terminating in an obtuse setiferous point. The inner ramus is small and scale-like, with 2 apical spines.

The telson (fig. 13) is comparatively small, and has the lateral lobes strongly diverging, each armed with only a single apical spinule.

The male does not differ from the female except by the anterior pairs of coxal plates being somewhat smaller. On the other hand, neither in the structure of the antennae nor in that of the gnathopoda or caudal appendages are there any differences to be detected, and this is probably the case with all the species belonging to this genus.

Occurrence. — Of this species solitary specimens were collected by Mr. Warpachowsky at 3 different Stations of the North Caspian Sea, the one (St. 58) located in the western part of that basin, north of the Tschistyi Bank, the 2nd (St. 61) occurring far north, at some distance outside the

Bay Bogutui Kultuk, and the 3rd (St. 63) lying somewhat farther south than the latter.

In the collection of Dr. Grimm this species is likewise represented only by quite solitary specimens collected in the southern and middle part of the Caspian Sea, the depth varying from 7 to 20 fathoms.

21. Niphargoides æquimanus, G. O. Sars, n. sp.

(Pl. XVIII, figs. 14—23).

Specific Characters. — Very much like the last described species, as to outer appearance, but of much smaller size. Cephalon considerably exceeding in length the 1st segment of mesosome, and having the lateral lobes rather produced and rounded at the tip. Anterior pairs of coxal plates comparatively smaller than in *N. quadrimanus*, and fringed with scattered bristles distally; 1st pair somewhat expanded in their outer part; 4th pair fully as broad as they are deep. Epimeral plates of metasome well developed and quite smooth. Urosome without any spines dorsally. Eyes comparatively small. Antennæ of a structure similar to that in *N. quadrimanus*, but with a less number of articulations in the flagella. Gnathopoda almost exactly alike both in structure and size, propodos in both pairs oblong quadrangular, with the palm much shorter than the hind margin. Pereiopoda resembling those in *N. quadrimanus*, except that the basal joint of last pair is still more expanded. Last pair of uropoda comparatively more elongated than in the said species. Telson with the lateral lobes scarcely diverging, each armed at the tip with 2 unequal spines. Length of adult male 5 mm.

Remarks. — This form is very nearly allied to *N. quadrimanus*, and may easily be confounded with it. On a closer examination, it is, however, found to differ, not only by its small size, but also in some structural details, especially in the structure of the gnathopoda and the shape of the basal joint of the last pair of pereiopoda. Finally the last pair of uropoda are more elongated, and the lateral lobes of the telson scarcely diverging.

Description of the male. — The length of an apparently adult specimen measures only 5 mm., and this form is accordingly much inferior in size to the other known species. The form of the body (see fig. 14) is rather slender and somewhat compressed, bearing on the whole a strong resemblance to that in *N. quadrimanus*.

The cephalon is almost as long as the first 2 segments of the mesosome combined, and has the lateral lobes rather prominent and narrowly rounded at the tip.

The anterior pairs of coxal plates are but little deeper than the corresponding segments, and are fringed on the distal edge with a restricted number of slender bristles. The 1st pair (see fig. 17) are somewhat expanded in their outer part, with the distal edge slightly curved. The 2 succeeding pairs are oval quadrangular in form, and obtusely truncated at the tip. The 4th pair are fully as broad as they are deep, and of the usual, irregular, angular shape.

The epimeral plates of the metasome are rather large, and without any trace of bristles. The last 2 pairs are nearly rectangular, whereas the 1st pair, as usual, are more rounded.

The urosome is comparatively stout, and has not any spines dorsally, the first 2 segments having only in the middle of the dorsal face a few small hairs.

The eyes are rather small and of an oval reniform shape, with dark pigment.

The antennæ (figs. 15, 16) exhibit a structure similar to that in *N. quadrimanus*, but have the flagella less fully developed, each being composed of only 7 articulations. The accessory appendage of the superior ones is scarcely half so long as the flagellum, and 5-articulate.

The gnathopoda (figs. 17, 18) are moderately strong, and, unlike what is the case in the other species, subequal, the propodos being in both pairs almost exactly alike both in size and shape. It is of an oblong quadrangular form, with the palm nearly transverse and much shorter than the hind margin. The spines issuing from the lower corner are less strong than in *N. quadrimanus*.

The pereiopoda resemble in their structure those in the said species. On closer comparison, however, some minor differences are to be found. Thus the basal joint of the antepenultimate pair (fig. 19) appears comparatively shorter in proportion to its breadth, and that of the last pair (fig. 20) has the posterior expansion still larger and more regularly rounded, with a smaller number of marginal bristles.

The 2 anterior pairs of uropoda (fig. 21) are likewise much of the same appearance as in *N. quadrimanus*, except that the rami want the lateral spine present in that species.

The last pair of uropoda (fig. 22) are still more slender than in the said species, the outer ramus being about 3 times as long as the basal part. It has but very few marginal bristles, and, as in *N. quadrimanus*, 2 fascicles of spines on the outer edge. The inner ramus has but a single apical spinule.

The telson (fig. 23) differs from that in the said species in having the lateral lobes comparatively broader and not at all diverging, each being armed at the tip with 2 unequal spinules.

Occurrence. — Of this species only 3, partly defective specimens were collected by Mr. Warpachowsky at Stat. 53, occurring north of the island of Kulaly.

In the collection of Dr. Grimm there is a single specimen, which was taken in the middle part of the Caspian Sea, near the western coast, from a depth of 10 fathoms.

Gen. 6. **Pandorites**, G. O. Sars.

Syn.: *Pandora*, Grimm.

Generic Characters. — Body but little compressed, and quite smooth above. Coxal plates of moderate size: 1st pair the smallest; 4th pair but slightly emarginated posteriorly. Epimeral plates of metasome well developed. Urosome short and stout. Eyes placed close to the lateral lobes of the cephalon. Antennae rather slender, but not much elongated, equal-sized, the superior ones with an accessory appendage. Oral parts normal. Gnathopoda very unequal, and of the same structure in the 2 sexes; the anterior ones of normal appearance, the posterior ones, however, peculiarly developed and rather powerful, resembling those in the genus *Gammaracanthus*, the propodos being greatly expanded distally, with the palm arcuate and having below a particularly long and slender spine. Pereiopoda not much elongated, and of normal structure, basal joint of last pair lamellarly expanded. Last pair of uropoda small. Telson likewise small and cleft to the base.

Remarks. — This genus has been established by Dr. Grimm to include a rather peculiar Gammarid from the Caspian Sea to be described below. But as the name he proposes, *Pandora*, has been used long ago, and as also the derivations *Pandorina* and *Pandorella* have been appropriated in Zoology, I propose to change the name to *Pandorites*. Besides the typical species, *P. podoceroides*, Dr. Grimm refers another form to the same genus under the name of *P. cocca*. But this form differs essentially both in the structure of the antennae and gnathopoda, and cannot therefore in my opinion be regarded as congeneric. The specimens of the latter form contained in the collection of Dr. Grimm and taken from the very considerable depth of 108 fathoms, would all seem to be still immature.

22. **Pandorites podoceroides**, Grimm, MS.

(Pl. XIX).

Specific Characters. — Body rather slender, with evenly rounded back, and exhibiting in its outer appearance some resemblance to that in the species of the genus *Podocerus*. Cephalon with the lateral lobes rather projecting and evenly rounded at the tip, postantennal corners produced to an acute point. Anterior pairs of coxal plates considerably deeper than the corresponding segments, and but sparingly setous; 1st pair much smaller than the others, and somewhat tapering distally; 4th pair rather broad, with the infero-posteal corners angularly produced. Last pair of epimeral plates of metasome almost rectangular. Urosome short and stout, with a few small hairs and spinules dorsally. Eyes of moderate size and oval in form, being placed just within the edges of the lateral lobes of the cephalon. Superior antennæ about twice the length of the cephalon, joints of the peduncle successively diminishing in size, flagellum nearly as long as the peduncle, accessory appendage comparatively small and 4-articulate. Inferior antennæ with the last 2 joints of the peduncle simple cylindric, flagellum about half the length of the peduncle. Anterior gnathopoda moderately strong and rather densely setous, propodos obpyriform, with the palm oblique and imperfectly defined below. Posterior gnathopoda much larger and rather elongated, with only scattered small bristles, basal joint subfusiform, the 3 succeeding ones comparatively small and narrow, propodos extremely large and gradually expanded distally, palm obliquely arcuate and defined below by a very slight angle, dactylus long and falciform. The 2 anterior pairs of pereiopoda of moderate size and rather densely setous: the 3 posterior pairs slightly increasing in length and comparatively strongly built, basal joint of last pair large and lamellar, its posterior expansion terminating below in a broadly rounded lobe, and having the edge smooth. Last pair of uropoda extremely small, outer ramus scarcely longer than the basal part and having the terminal joint quite rudimentary, inner ramus scale-like, with a single apical seta. Telson small, lateral lobes not diverging, each with a single apical spine. Length of adult female 11 mm., of male 13 mm.

Remarks. — This is the only as yet known species of the genus, the form named by Dr. Grimm *Pandora cocca* being, as above stated, not congeneric.

Description of the female. — The length of fully adult, ovigerous specimens is about 11 mm.

The form of the body (see fig. 1) is somewhat slender and scarcely at all compressed, the back being broadly rounded and quite smooth throughout. On the whole it bears an unmistakable resemblance to that in some species of the genus *Podocerus*, or rather *Ischyrocerus;* hence the specific name proposed by Dr. Grimm.

The cephalon is not fully so long as the first 2 segments of the mesosome combined, and forms (see fig. 2) a slight angular projection in front.

The lateral lobes considerably project between the bases of the 2 pairs of antennæ, and are quite evenly rounded at the tip. The postantennal corners are produced to an acuminate, anteriorly curving process.

The anterior pairs of coxal plates are considerably deeper than the corresponding segments, and are, excepting the 1st pair, but very sparingly setous at the distal edge. The 1st pair (see fig. 12) are much smaller than the others and somewhat tapered distally, with the tip obliquely rounded and fringed with a number of rather elongated setæ. The 2 succeeding pairs (see figs. 13, 14) are comparatively broad, and subrhomboidal in shape, with the terminal edge obtusely rounded. The 4th pair (see fig. 16) are still somewhat broader and but very slightly emarginated posteriorly, with the posterior expansion not, as usual, truncated, but terminating in a single angular corner.

The 3 posterior pairs of coxal plates successively decrease in size, the antepenultimate pair (see fig. 17) being considerably larger than the other 2, though not nearly so deep as the anterior pairs.

The epimeral plates of the metasome are of moderate size and perfectly smooth. The 1st pair, as usual, exhibit a rounded form, whereas the 2 succeeding pairs are almost rectangular.

The urosome is comparatively short and stout, with a few small hairs and spinules dorsally.

The eyes (see fig. 2) have a somewhat unusual position, being placed close to the edges of the lateral lobes of the cephalon, and also by this character the present form acquires some habitual resemblance to the species of the genus *Podocerus*. They are of moderate size and oval in form, with the visual elements well developed and the pigment of a dark hue.

The superior antennæ (fig. 3) are rather slender, but not very much elongated, scarcely exceeding twice the length of the cephalon. The 1st joint of the peduncle is much the largest, being fully as long as the other 2 combined, and, like the latter, is provided at the end with slender bristles. The 3rd joint is rather small, about half the length of the 2nd. The flagellum is nearly as long as the peduncle, and composed of 7 articulations. The accessory appendage is rather small, being about $\frac{1}{3}$ as long as the flagellum, and 4-articulate.

The inferior antennæ (fig. 4) are about the length of the superior, and of quite normal structure, being, as the latter, clothed with scattered fascicles of slender bristles. The 2 outer joints of the peduncle are simple cylindric, and successively diminish both in length and breadth. The flagellum does not attain the length of those joints combined, and is composed of 5 rather slender articulations.

The oral parts (figs. 5—11) are of quite normal structure, and need not therefore be described in detail.

The anterior gnathopoda (fig. 12) likewise exhibit quite a normal appearance, being moderately strong and rather densely setiferous. The propodos is somewhat tumid, and of an ovate, or rather obpyriform shape, with the palm not defined below by any distinct angle, but carrying at the junction with the hind margin the usual spines.

The posterior gnathopoda (fig. 13), on the other hand, are quite unlike the anterior, and of a rather peculiar structure, strongly reminding of that characteristic of the genus *Gammaracanthus*. They are much larger than the anterior ones and considerably elongated, being also much less densely setiferous. The basal joint is large and dilated on the middle, exhibiting a somewhat fusiform shape, and is filled with strong muscles moving the outer part of the leg. The 3 succeeding joints are comparatively small and narrow, the carpal one being produced below to a short and narrow setiferous lobe. The propodos is exceedingly large, and gradually expands distally, acquiring thereby a somewhat flattened shape. The palm is longer than the hind margin and obliquely curved, its edge being sharpened and fringed with a regular row of small bristles. The defining angle is very slight, and is (see fig. 14) armed with 3 comparatively short spines, behind which there are 2 or 3 fascicles of comparatively short bristles. Inside the angle, as in most other Gammaridæ, 2 juxtaposed spines occur, the outer of which is exceedingly slender and elongated. Between these 2 spines and those of the defining angle the tip of the slender, falciform claw is received when impinged.

The anterior pairs of pereiopoda (figs. 15, 16) do not exhibit any essential peculiarity in their structure. They are rather densely setous and somewhat unequal in size, the 1st pair (fig. 15) being the larger.

The 3 posterior pairs of pereiopoda (figs. 17—19) are comparatively strongly built and not much elongated, being provided in their outer part with fascicles of slender bristles. The antepenultimate pair (fig. 17) are, as usual, somewhat shorter than the other 2, and have the basal joint oval quadrangular in form, with the anterior edge slightly curved and throughout provided with fascicles of slender bristles. The basal joint of the penultimate pair (fig. 18) is more elongated and somewhat narrowed distally, with 4 fascicles of bristles on the outer part of the anterior edge. The last pair (fig. 19) are distinguished by the large size of the basal joint, which forms posteriorly a broad lamellar expansion terminating below in a rounded lobe. The edges of the expansion are obscurely serrate, but without any trace of bristles. On the other hand the anterior edge of this joint is very densely setiferous in its outer part.

The 2 anterior pairs of uropoda (figs. 20, 21) are comparatively strongly built, with the rami subequal and armed at the tip with blunt spines. In the 1st pair (fig. 20) each of the rami has besides a lateral spine, whereas in the 2nd pair (fig. 21) this is only the case with the inner ramus.

The last pair of uropoda (fig. 22) are extremely small, reaching but little beyond the others. The outer ramus is scarcely longer than the basal part, and has one lateral and 2 apical spines, but no lateral setæ. The terminal joint of this ramus is so very minute as easily to escape attention, forming only a diminutive nodule tipped with a few hair-like bristles. The inner ramus is of the usual scale-like character and provided with a single apical bristle.

The telson (fig. 23) is likewise unusually small, scarcely reaching beyond the basal part of the last pair of uropoda. It is divided by a deep and narrow cleft into 2 halves, each of which carries on the somewhat truncated tip a single spine accompanied by a small hair.

The adult male (fig. 24) is somewhat larger than the female, attaining a length of about 13 mm., and has the metasome somewhat more fully developed, but is otherwise of a very similar appearance. In the structure of the antennæ, no other difference is to be found than that the flagella have a somewhat greater number of articulations. Also the gnathopoda exhibit a structure very similar to that in the female, though the posterior ones appear somewhat larger.

Occurrence. — Of this interesting form a few specimens were collected by Mr. Warpachowsky at Stat. 63, in the eastern part of the North Caspian Sea.

The collection of Dr. Grimm contains solitary specimens, derived from 4 different Stations, one of which is located in the southern part, the other 3 in the middle part of the Caspian Sea, the depth ranging from 7 to 48 fathoms.

Fam. COROPHIIDÆ.

Gen. **Corophium**, Latr.

Remarks. — This genus, as is well known, has hitherto been regarded as exclusively marine, no species having ever been found in fresh water; and even in brackish water it is rather seldom to meet with these peculiar Amphipods, which on the whole would seem to be restricted to the open Ocean coasts, where the water is very salt. From the Black Sea only 3 species are recorded by Mr. Sowinsky, and one of these, described as

C. longicorne var. *lævicorne*, is evidently not at all a Corophium, but a true *Siphonœcetes*. The remaining 2 species are *C. Bonelli* Edw. and *C. crassicorne* Bruzel., both known also from the European coasts, and it is most probable, that the form recorded by Dr. Marcusen as *C. bidentatum* is identical with the last named species, in which case only 2 species are met with in the Black Sea. It was therefore rather unexpected to find this genus very abundantly represented in the Caspian Sea, both as to species and individuals. On a closer examination of the rich material of Corophians collected by Mr. Warpachowsky, I have been enabled to distinguish no less than 6 different species, all of which are new to science, exhibiting well marked differences from those earlier known. It will be shown below that the species are rather easily distinguishable especially by the structure of the inferior antennæ, those of the male sex particularly exhibiting the distinguishing characters very clearly pronounced. The Corophians are known to be chiefly littoral and sublittoral in their occurrence, living partly among algæ, partly at muddy bottom, and in both cases constructing for themselves abodes of mud or other material for dwelling in. The same habits are also to be stated for the Caspian species, and their muddy tubes are often found together with the specimens, in several cases containing within them the animal in its original position.

23. Corophium nobile, G. O. Sars, n. sp.

(Pl. XX and XXI).

Specific Characters. — Cephalon angularly produced between the bases of the superior antennæ, lateral corners narrowly rounded. First pair of coxal plates densely clothed with slender, partly ciliated setæ. All the segments of urosome distinctly defined. Superior antennæ very slender and elongated, exceeding in male half the length of the body; peduncle, especially in the male, densely setiferous, its 1st joint having in both sexes 2 distant spines on the lower edge, flagellum in male exceeding the length of the peduncle. Inferior antennæ in male very strongly developed, equalling in length about $^3/_4$ of the body, penultimate joint of the peduncle rather large and tumid, being produced at the end posteriorly to 2 somewhat diverging unguiform projections, the outer of which is the larger, last joint somewhat shorter than the penultimate one, and having above the middle a short spiniform prominence, but no spine at the end. Inferior antennæ in female much less strong than in male, but of a similar structure, though the projections of the penultimate peduncular joint are smaller and less divergent. Gnathopoda of the structure characteristic of the genus. Anterior pairs of pereiopoda comparatively slender, with the meral joint not much expanded, and in male densely clothed with slender bristles anteriorly. Last pair of pereiopoda very much elongated, exceeding half the length of the body, basal joint rather expanded and, as usual, provided on both edges with a double row of partly ciliated setae, outer joints very slender and narrow. The 2 anterior pairs of uropoda strongly built and densely spinous; last pair small, with the terminal joint oval lamelliform and densely setiferous. Telson about twice as broad as it is long, and provided at the end above with 2 lamelliform crests, each divided into 4 recurved teeth, tip transversely truncate. Length of adult female 10 mm., of male 11 mm.

Remarks. — This is the largest and finest of the Caspian species, and is easily recognizable by the slender and elongated superior antennae, and by the structure of the inferior ones. Moreover the comparatively slender form of the anterior pairs of pereiopoda may serve to easily distinguish this species from the other Caspian forms.

Description. — The length of adult, ovigerous females is about 10 mm., that of males 11 mm., and this form accordingly grows to a considerably larger size than any of the other known species.

The form of the body (see Pl. XX, figs. 1 and 2, Pl. XXI, fig. 1) is that characteristic of the genus, being subdepressed, with the back broadly vaulted, and the lateral parts of the segments extended horizontally. As seen from above (Pl. XXI, fig. 1), the body appears nearly of equal breadth throughout, exhibiting a somewhat linear form.

The cephalon is broad, subdepressed, and exceeds somewhat in length the first two segments of the mesosome combined. The frontal edge is (see Pl. XXI, fig. 1) angularly produced in the middle, and the lateral corners project as narrowly rounded lobes between the bases of the 2 pairs of antennae. Behind these lobes the lateral edges of the cephalon form (see Pl. XX, figs. 1 and 2) a broad emargination encircling the base of the inferior antennae.

The coxal plates are, as in the other species of the genus, very small and scale-like. The 1st pair (see Pl. XX, fig. 12) are, however, somewhat more fully developed, being produced anteriorly to a narrowly rounded lobe clothed with numerous slender, anteriorly curving setae, some of which are finely ciliated. The 3 posterior pairs are slightly bilobed, with the anterior lobe the larger.

The epimeral plates of the metasome are rather shallow, and all of them obtusely rounded at the lateral corners, their edges being densely fringed with ciliated bristles. Those of the last pair are much larger than the others in accordance with the greater development of the corresponding segment.

The urosome (Pl. XXI, fig. 9) is short and stout, much depressed, and divided into 3 distinctly defined segments rapidly diminishing in size.

The eyes are small, rounded, and located at the bases of the lateral lobes of the cephalon. The ocular pigment is of a dark hue, but the visual elements would seem to be less perfectly developed.

The superior antennæ (Pl. XX, fig. 3, Pl. XXI, fig. 2) are very slender, and somewhat more elongated in the male than in the female, considerably exceeding half the length of the body in the former. The peduncle is densely setiferous, especially in the male, and, as usual, is composed of 3 distinctly defined joints, the 1st of which is much the largest, though not fully so long as the other 2 combined. In both sexes this joint is armed on the lower edge with 2 distant spines. The 3rd joint is scarcely more than half as long as the 2nd and very narrow. The flagellum is extremely slender, filiform, equalling in the female about the peduncle in length, in the male considerably longer, and divided into about 20 short articulations.

The inferior antennæ (Pl. XX, fig. 4, Pl. XXI, fig. 3) are in both sexes subpediform, but much larger in the male than in the female, exceeding in the former $3/4$ of the length of the body. The peduncle is only composed of 4 joints, the first 2 being fused together. The penultimate joint is much the largest, and especially in the male very much tumefied, exhibiting a somewhat fusiform shape. It is in both sexes produced at the end posteriorly to 2 strong unguiform projections, the outer of which is the larger. These projections are, however, much coarser and more divergent in the male than in the female (comp. Pl. XX, fig. 4 and Pl. XXI, fig. 3). The last peduncular joint, which is very movably articulated to the penultimate one, is somewhat shorter than the latter and much narrower, being sublinear in form. It is, like the preceding joints, provided inside with fascicles of slender bristles, and has the posterior edge produced above the middle to a short and stout, somewhat recurved projection, which is received between the projections of the preceding joint, when the outer part of the antenna is bent in against the inner. The flagellum is not fully so long as the last peduncular joint, and is composed of 3 articulations, the last 2 of which, however, are very small. It is densely clothed on both edges with fascicles of slender bristles. At the tip it has a dense brush of very delicate bristles, between which, on close examination, 2 short curved hooks are found to project, both issuing from the extremely small terminal joint (see Pl. XXI, fig. 4).

The buccal area (see Pl. XX, figs. 1 and 2) is not much protuberant, and partly covered by the 1st pair of coxal plates. The oral parts, though exactly agreeing with those in the other species of the genus, may here be described in detail, as they in some points differ rather markedly from those in the *Gammaridæ*.

The anterior lip (Pl. XX. fig. 5) is broadly quadrangular in form, with a dentiform projection in front. The terminal edge is very slightly emarginated and finely ciliated.

The posterior lip (fig. 6) is rather large, with distinctly developed inner lobes. The outer lobes are narrowly rounded at the tip, which is edged with delicate cilia, and project outside to a narrow lappet.

The mandibles (figs. 7. 8) are short and stout, with a well-developed molar expansion. The cutting edge is in both mandibles divided into 2 superposed lamellæ, the outer of which is distinctly dentated, whereas the inner is very narrow, spiniform, especially on the right mandible. Behind the cutting edge there occur on the left mandible 3, on the right only 2 ciliated spines. The palp (see fig. 7) is very small, and composed of only 2 joints of about equal length, and generally forming together a strong geniculate bend. The 1st joint has at the end a single spiniform seta, and a similar, though somewhat more slender seta issues from the tip of the very narrow, conically tapering terminal joint.

The 1st pair of maxillæ (fig. 9) would seem wholly to want the basal lobe. The masticatory lobe is narrowly truncated at the end, which carries several slender spines. The palp is well developed and biarticulate, with the terminal joint somewhat expanded distally and armed at the tip with a number of small spines.

The 2nd pair of maxillæ (fig. 10) are rather fully developed, being scarcely smaller than the 1st pair. The inner lobe is somewhat curved and narrowed distally, having at the tip a dense clothing of small spines and along the inner edge a regular row of slender, ciliated setæ. The outer lobe is considerably larger than the inner and somewhat expanded distally, with a dense brush of slender spines on the obtusely truncated tip.

The maxillipeds (fig. 11) exhibit all the pertaining parts well developed. The basal lobes are of a somewhat unusual form, being conically tapered distally, and having along the inner edge a row of slender curved setæ. The masticatory lobes are very much elongated, narrow linguiform in shape, and fringed along the inner edge with numerous very delicate bristles. The palps are slender, and rather densely setiferous, with the joints somewhat laminar. The last joint is comparatively small and narrow, and the dactylus is extremely minute, knob-shaped, and setous at the tip.

The anterior gnathopoda (fig. 12) are comparatively slender and feeble in structure. The basal joint is rather narrow, though gradually somewhat widening distally. The ischial joint is short and thick, and carries below a dense transverse row of very slender anteriorly curving setæ. The meral joint is so very minute as easily to escape attention. The carpus, on the other hand, is very large and compressed, almost fusiform in outline, and is very densely setiferous, especially on the lower edge. The propodos is somewhat shorter than the carpus, slightly curved, and rather narrow, though gradually widening distally, being densely clothed anteriorly with slender bristles, partly arranged in transverse rows. The palm is very short and transverse, being defined below by a distinctly projecting corner; its edge is minutely spinulose. The dactylus is comparatively slender, and extends considerably beyond the defining corner of the palm, when closed.

The posterior gnathopoda (fig. 13) are more strongly built than the anterior, and of a very different structure. The basal joint is rather thick and not much elongated, being firmly connected with the extremely short, nearly band-shaped ischial joint. The meral joint is peculiarly developed, being produced along the lower side of the carpus to a broad, lamellar expansion, which is firmly connected to the latter, though defined by a distinct suture. The expansion, which extends until the end of the carpus, is fringed with a double row of exceedingly long and slender setæ, which are finely ciliated and curved anteriorly, forming together a broad fan. The propodos is very narrow and elongated, sublinear in form, and clothed on both edges with fascicles of slender bristles. It projects at the end below the dactylus to an acute corner; but no distinct palm is present. The dactylus is slender and curved, being armed on the concave edge with 4 strong secondary denticles.

The 2 anterior pairs of pereiopoda (Pl. XXI, fig. 5) are exactly alike both in size and structure, and are rather slender, compared with those in the other species. The basal joint is slightly expanded and, as in the other species, contains a glandular mass, which probably serves to secrete a viscid fluid to be used for constructing the dwelling tubes. The meral joint is about the length of the last 2 combined, and is not much expanded, terminating at the end anteriorly in an obtuse corner. Anteriorly this joint is in the male clothed with slender diverging bristles, forming a very dense brush. The propodal joint is very narrow and conically tapering, with scattered small bristles on the edge and at the tip. The dactylus is about the length of that joint and very slender, terminating in a sharp point.

The 2 succeeding pairs of pereiopoda (figs. 6 and 7) are comparatively short and stout, and of essentially the same structure, though somewhat unequal in size, the antepenultimate pair (fig. 6) being considerably shorter than the penultimate one (fig. 7). In both pairs the basal joint is rather expanded and of an oval fusiform shape, but in the antepenultimate pair its posterior edge is nearly straight and perfectly smooth, whereas in the penultimate pair it is arched and fringed with a number of ciliated setæ. The meral joint gradually widens distally and is obliquely truncated at the end, with the anterior corner more prominent than the posterior. The carpal joint is considerably smaller, and likewise obliquely truncated at the end, but in an inverted manner, the posterior corner being the more prominent. On the outer side of this joint there are 2 oblique rows of strong curved spines, the lower row, terminating at the posterior corner, containing 6 spines successively increasing in length distally. The propodal joint is very narrow, sublinear, and much longer than the carpal one. The dactylus is comparatively short and strongly curved, being more or less extended outwards, for which reason it often appears inverted. Both those pairs of legs are generally found to be strongly reflexed, with their outer part extended laterally (see fig. 1), and it is most likely that they are of essential service in affixing the animal within its tube.

The last pair of pereiopoda (fig. 8) exhibit an appearance very different from that in the 2 preceding pairs. They are very slender and elongated, considerably exceeding half the length of the body, and are generally extended straight backwards. The basal joint is lamellarly expanded and broadly oval in form, though somewhat tapering distally. It is fringed on both edges with numerous slender plumose setæ, arranged in a double row, those of the anterior edge being generally curved downwards. The outer joints are very narrow and increase somewhat in length, the propodal one being the longest. They are clothed with fascicles of slender bristles, those issuing from the end of the joints being particularly elongated. The dactylus is of moderate length, somewhat curved, and terminates in a very acute point.

The branchial lamellæ (see fig. 5 and 6) are simple, oblong oval in form, and only present at the base of the 4 anterior pairs of pereiopoda.

The incubatory lamellæ (see Pl. XX, fig. 1 and 13) are present at the base of all the legs, except the anterior gnathopoda and the last pair of pereiopoda. They are narrow linguiform in shape so as not to fit together with their edges. As they, however, are all round fringed with strong incurved setæ, the ova in the marsupial pouch are by these means securedly kept in place.

The pleopoda (Pl. XX, fig. 14) are distinguished by the unusual development of the basal part, which is produced inside to a very large and broad, sublaminar expansion, into which a bundle of strong muscular fibres are seen to pass. Inside the obtuse tip of this expansion 2 peculiarly constructed spines are found to be secured, being placed close together and provided with small recurved hooks (see fig. 15). By the aid of these spines, which meet the corresponding ones on the adjacent pleopod, both are bound together, so as only to be admitted to move simultaneously. The rami, which issue close together from the outer corner of the basal part, are turned obliquely inwards, and are divided into numerous short articulations, each carrying a pair of long natatory setæ.

The 2 anterior pairs of uropoda (Pl. XX, figs. 9, 10, 11) are essentially of the same structure, though rather different in size, the 1st pair being much the larger. They are rather strongly built, with both the basal part and the rami coarsely spinous. In the 1st pair (fig. 10) the basal part is nearly twice as long as the rami, and armed in the distal part of the inner edge with 4 very strong spines, the outer edge being minutely spinulose throughout. The rami are subequal and narrowly rounded at the tip, each carrying from 17 to 18 spines, which are more densely crowded on the outer edge, those issuing from the tip being longer than the others. In the 2nd pair (fig. 11) the basal part is but little longer than the rami, and, like the latter, has a smaller number of spines.

The last pair of uropoda (fig. 12) are very unlike the preceding ones, and rather small, scarcely reaching beyond the basal part of the 2nd pair. They are simple, not biramous, being composed of 2 joints of about equal size, the latter of which is somewhat lamellar and oval in form, being clothed at the obtusely rounded tip with a dense brush of slender bristles.

The telson (fig. 13) is nearly twice as broad as it is long, and somewhat narrowed distally. The tip is entire and almost transversely truncated, being flanked on each side by a projecting vertical crest, which is divided into 4 small recurved teeth, best seen in a lateral view of the animal (see fig. 14). No doubt, this peculiar structure of the telson, which seems to be common to all the species of the genus, may stand in some connexion with the tubicolous nature of the animal, serving in all probability to affix the animal within its tube.

Colour. — As in most other species, the body is ornamented with a dark brown pigment, which is pretty well observable even in specimens for a long time preserved in spirit. This pigment is (see Pl. XX, figs. 1 and 2, Pl. XXI, fig. 1) chiefly restricted to the dorsal face of the animal, forming on the cephalon a distinctly defined dark longitudinal band, which expands in front so as nearly to occupy the whole breadth of the cephalon (see Pl. XXI, fig. 1). On the mesosome the pigment forms in each segment 2 more or less distinct transverse bands, which are confluent in the middle of the dorsal face. In the metasome and urosome the pigment is generally more irregularly distributed. Moreover, some of the appendages of the body are more or less distinctly pigmented; and especially the inferior antennæ in the male show a rather peculiar arrangement of the pigment, as shown in fig. 2 on Pl. XX and fig. 1 on Pl. XXI.

Occurrence. — This pretty species has been collected by Mr. Warpachowsky in 9 different Stations of the North Caspian Sea, though in none of them occurring in any considerable number. Of the Stations 6 (St. 17, 23, 24, 26, 52, 53) are distributed in the tract north of the peninsula Mangyschlak, the other 3 (St. 61, 63, 64) in the northern and eastern part of the basin.

In the collection of Dr. Grimm this species is also represented, having been collected in several places both of the southern and middle part of the Caspian Sea, at a depth ranging from 6 to 40 fathoms.

24. Corophium chelicorne, G. O. Sars, n. sp.

(Pl. XXII).

Specific Characters. — Frontal edge of cephalon not produced in the middle, lateral lobes narrowly rounded. First pair of coxal plates with only 3 slender bristles at the tip. The last 2 segments of urosome less distinctly defined. Superior antennæ but sparingly setous, and in female scarcely exceeding $\frac{1}{3}$ of the length of the body, 1st joint of the peduncle about the length of the other 2 combined, and in female armed below with about 7 spinules, in male without any such spinules, 2nd joint in male considerably longer than in female, flagellum in both sexes shorter than the peduncle. Inferior antennæ very strongly built, especially in the male, penultimate joint of the peduncle exceedingly large and produced at the end posteriorly to a very prominent, acuminate, thumb-like projection having inside a small secondary tooth; last peduncular joint scarcely more than half as long as the preceding one, and armed below the middle with a short recurved projection, being moreover produced at the end to a strong spiniform process, which crosses the end of the thumb-like projection when the joint is incurved, thereby giving these antennæ a pronounced cheliform character; flagellum about the length of the last peduncular joint, and of the usual structure. Gnathopoda scarcely differing in their structure from those in the preceding species. Anterior pairs of pereiopoda somewhat stronger, but rather much elongated, with the meral joint longer than the last 2 combined, and gradually widening distally, anterior edge scarcely setous. Last pair of pereiopoda somewhat shorter and less slender than in the preceding species, otherwise of a very similar appearance. The 2 anterior pairs of uropoda with the rami spinous only at the tip and the outer edge. Last pair of uropoda and telson nearly as in *C. nobile*. Length of adult female 7 mm., of male 8 mm.

Remarks. — The present species is easily recognizable by the peculiar structure of the inferior antennæ, which exhibit, as it were, a cheliform character, on account of the great development of the projection issuing from the penultimate joint of the peduncle, which forms a sort of thumb, against which another spiniform process originating from the last peduncular joint, admits of being impinged; hence the specific name.

Description. — The length of fully adult ovigerous females is about 7 mm., that of males 8 mm., and this species is accordingly somewhat inferior in size to the preceding one.

The form of the body (see figs. 1 and 5) appears on the whole somewhat less slender than in *C. nobile*, but is otherwise rather similar.

The cephalon is about the length of the first 2 segments of the mesosome combined, and has the frontal edge not at all produced in the middle being only slightly arcuate (see fig. 2). The lateral lobes are narrowly rounded and not very prominent.

The coxal plates are of exactly the same shape as in the preceding species, but the 1st pair (see fig. 8) have only 3 slender bristles on the tip and a few small hairs on the anterior edge.

The epimeral plates of the metasome likewise agree with those in the said species.

The urosome (fig. 14) exhibits the usual short, flattened form, and has the 1st segment very distinctly defined. On the other hand is the line of demarcation between the 2 other segments far less distinct, though they are not perfectly fused together, as is the case in some other known species.

The eyes are very small and rounded, with dark pigment.

The superior antennæ are (see figs. 1 and 5) comparatively shorter than in the preceding species, and in the female scarcely exceed $\frac{1}{3}$ of the length of the body. In the male they are, as usual, somewhat more elongated, though not nearly to such an extent as in the male of *C. nobile*. They are in both sexes but sparingly supplied with bristles, and have the 1st joint of the peduncle about as long as the other 2 combined. In the female this joint (see fig. 3) is armed below with several acute spinules, generally 7 in number, whereas in the male (see fig. 6) no trace of such spinules are found. In the latter the 2nd peduncular joint is considerably more elongated than in the female, being more than twice as long as the 3rd. The flagellum is in both sexes shorter than the peduncle, and is composed in the female of 10, in the male of 15 articulations.

The inferior antennæ (figs. 4 and 7) are in both sexes very strongly built, though, as usual, much larger in the male than in the female, equalling in the former ²/₅ of the length of the body. The penultimate joint of the peduncle is exceedingly large and tumid, and is produced at the end posteriorly to a very prominent, thumb-like projection terminating in an acuminate point, and having inside a well marked secondary tooth. This projection is comparatively more strongly developed in the male than in the female (comp. figs. 4 and 7), but in both sexes extend until the end of the last peduncular joint. The latter exhibits the usual cylindric shape, and is scarcely more than half as long as the penultimate joint. It has inside, somewhat below the middle, a short and stout recurved prominence, and is moreover produced at the end to a strong spiniform process. When the joint is bent in, this process crosses the tip of the thumb-like projection of the preceding joint, whereby the antenna acquires a pronounced cheliform character (see fig. 17). The flagellum is about the length of the last peduncular joint, and of same structure as in the preceding species.

The gnathopoda (figs. 8—9) agree nearly exactly in their structure with those in the said species, and need not therefore be described in detail.

The 2 anterior pairs of pereiopoda (fig. 10) appear somewhat more strongly built, though they are rather elongated. The meral joint is somewhat longer than the last 2 combined, and gradually expands distally, terminating in front in an obtuse, setiferous prominence. The anterior edge of this joint is in both sexes nearly quite smooth. The carpal joint is rather short, and the propodal one less slender than in *C. nobile*. The dactylus is not fully so long as the propodal joint, and very acute.

The 2 succeeding pairs of pereiopoda (figs. 11, 12) do not exhibit any essential difference from those in the preceding species.

The last pair of pereiopoda (fig. 13) are likewise of a very similar structure, though being perhaps not quite so slender as in *C. nobile*. Of the outer joints, the propodal one is particularly elongated, being nearly twice as long as the carpal one.

The uropoda (see fig. 14) agree on the whole with those in the preceding species, except that the rami of the 2 anterior pairs are spinous only at the tip and the outer edge.

The telson (ibid.) would likewise seem to be constructed in the same manner as in that species.

Also the pigmentation of the body resembles that observed in *C. nobile*.

Occurrence. — This species has been collected by Mr. Warpachowsky at no less than 10 different Stations of the North Caspian Sea. Of the Stations one (St. 6) is located near the western coast, at the entrance of the Bai Agrachansky, 3 others (St. 53, 54, 56) north and west of the island Kulaly, the remaining 6 Stations (St. 61, 63, 64, 66, 69, 86) in the eastern part of the basin. At two of the Stations (St. 63 and 69) it occurred in great abundance.

The species is also rather abundantly represented in the collection of Dr. Grimm, having been collected in several localities both of the southern and middle part of the Caspian Sea, the depth ranging from 6 to 44 fathoms.

25. Corophium curvispinum. G. O. Sars, n. sp.

(Pl. XIII, fig. 1—9).

Specific Characters. — Frontal edge of cephalon slightly angular in the middle, lateral lobes rather prominent and narrowly rounded. First pair of coxal plates with 3 slender bristles at the tip. Urosome with the 2 outer segments less distinctly defined. Superior antennæ in female comparatively short, not attaining $\frac{1}{3}$ of the length of the body, and but sparingly setous, 1st joint of the peduncle with 4—5 spinules below, flagellum shorter than the peduncle: those in male much more fully developed, and having the peduncle densely setiferous below, its 2nd joint much elongated, being fully as long as the 1st, flagellum scarcely exceeding the length of the 2 outer peduncular joints combined. Inferior antennæ much larger in male than in female, attaining in the former almost the whole length of the body, penultimate joint of the peduncle gradually widening distally, and produced at the end posteriorly to a strongly incurved spiniform projection, at the base of which is a short, slightly bilobed expansion; last peduncular joint nearly as long as the penultimate one, and having near the base inside a short recurved prominence, but no spine at the end; flagellum shorter than the last peduncular joint. The 2 anterior pairs of perciopoda comparatively short and stout, with the meral joint much expanded. Last pair of perciopoda moderately slender and of the usual structure. Uropoda and telson nearly as in *C. chelicorne*. Length of adult female 6 mm., of male 7 mm.

Remarks. — As in the other species, the most prominent distinguishing character is also in this form the structure of the inferior antennæ, which is rather peculiar, and, as usual, more pronounced in the male than in the female. Moreover the structure of the superior antennæ in the male and that of the 2 anterior pairs of perciopoda will serve to easily distinguish this species from any of the 2 preceding ones.

Description. — The length of fully adult, ovigerous females does not exceed 6 mm., that of males being about 7 mm., and this form accordingly is still somewhat smaller than *C. chelicorne*.

The form of the body (see fig. 1) is on the whole much like that in the 2 preceding species, though perhaps a little more slender than in *C. chelicorne*.

The cephalon about equals in length the first 2 segments of the mesosome combined, and has the frontal edge slightly angular in the middle. The lateral lobes are rather prominent, and narrowly rounded at the tip.

The coxal and epimeral plates do not differ essentially from those in *C. chelicorne*, and the urosome (fig. 8) exhibits likewise a similar appearance to that in the said species, the last 2 segments being less sharply defined.

The eyes are small, and, as usual, placed at the bases of the lateral lobes of the cephalon.

The superior antennæ are rather different in the two sexes. In the female they are (fig. 2) comparatively short, scarcely attaining $^1/_3$ of the length of the body, and are rather sparingly setous. The 1st joint of the peduncle is about the length of the other 2 combined, and is armed below with 4—5 small spinules. The 2nd joint has a similar spinule in the middle of the posterior edge. The flagellum is shorter than the peduncle, and composed of about 9 articulations. In the male these antennæ (see figs. 1 and 4) are much more fully developed, and have the peduncle densely clothed with fascicles of slender bristles. The 2nd peduncular joint is considerably elongated, fully equalling in length the 1st one, but is, as usual, much narrower. The flagellum scarcely exceeds in length the last 2 peduncular joints combined, and is composed of about 12 articulations.

The inferior antennæ likewise exhibit a rather different appearance in the two sexes, being in the male (see fig. 1) much more fully developed than in the female (fig. 3), attaining in the former almost the length of the whole body. The penultimate joint of the peduncle gradually widens distally, and is produced at the end posteriorly (see fig. 5) into a strongly incurved spiniform projection, at the base of which is a small, slightly bilobed expansion. The last peduncular joint is rather elongated, being nearly as long as the penultimate one, but, as usual, much narrower, and of simple cylindric form. It is armed, at a short distance from the base inside, with a stout recurved prominence, but it has no spine at the end. The flagellum is shorter than the last peduncular joint, and exhibits the usual structure.

The gnathopoda scarcely differ in their structure from those in the 2 preceding species.

The 2 anterior pairs of pereiopoda (fig. 6) are, on the other hand, considerably shorter and stouter, with some of the joints lamellarly expanded. The basal joint is rather broad, with the anterior edge curved and fringed with about 10 slender setæ. The meral joint is considerably expanded, being almost as broad as it is long, and is setous on both edges. The last 2 joints are comparatively short, and the dactylus is fully as long as the propodal joint.

The last pair of perciopoda (fig. 7) exhibit the usual slender form, and are about half as long as the body.

The 2 anterior pairs of uropoda (see fig. 8) are rather short and stout, especially the 2nd pair (fig. 9), and the rami have a smaller number of spines than in the 2 preceding species.

The last pair of uropoda (see fig. 8) are somewhat narrower than in those species; otherwise of a very similar appearance. This is also the case with the telson.

The pigment of the body is arranged in a manner similar to that found in the 2 preceding species.

Occurrence. — This species, as the preceding one, has been collected by Mr. Warpachowsky at no less than 10 different Stations of the North Caspian Sea. Of these Stations, 2 (St. 2 and 50) are located in the western part of the basin, off the Tschistyi Bank, another (St. 21) at the point of the peninsula Mangyschlak, 4 others (St. 16, 17, 27, 52) in the neighbourhood of the islands Kulaly and Morskoy, and the remaining 3 (St. 32, 55, 56) between these islands and the opposite western coast. At Station 32 and 55 the species occurred rather plentifully.

The species is also represented in the collection of Dr. Grimm, having been taken in the Bays of Baku and Schachowaja from the shore to 5 fathoms. Moreover, numerous specimens of a *Corophium*, extracted from the intestine of an *Accipenser stellatus* and preserved in the same collection, have, on a closer examination, turned out to belong exclusively to this species.

26. Corophium robustum [1]), G. O. Sars, n. sp.

(Pl. XXIII, figs. 10—16).

Specific Characters. — Body rather robust, with broad flattened back. Frontal edge of cephalon very slightly angulated in the middle, lateral lobes narrowly rounded. Coxal plates and urosome about as in the 2 preceding species. Superior antennæ in female comparatively short, not attaining $\frac{1}{3}$ of the length of the body, in male somewhat more elongated and having the peduncle densely clothed with bristles, 1st joint of the peduncle in female with 3 small spinules below, 2nd joint in both sexes shorter than

---

[1]) In the plate this species is named *C. bidentatum*; but as this name has been previously used by Dr. Marcusen for an apparently different species from the Black Sea, I have changed the name to *robustum*.

the 1st, flagellum not nearly attaining the length of the peduncle. Inferior antennæ in both sexes very strongly built, though, as usual, somewhat larger in male than in female; penultimate joint of the peduncle large and tumid, being produced at the end posteriorly to a moderately long and but slightly curved spiniform projection, at the base of which, as in *C. curvispinum*, there is a short bilobular expansion; last peduncular joint much shorter than the penultimate one, and having somewhat above the middle posteriorly a short recurved prominence, end of the joint produced to a strong spiniform process; flagellum shorter than the last peduncular joint. The 2 anterior pairs of pereiopoda resemble those in *C. curvispinum*, though they are somewhat more elongated; meral joint rather much expanded and densely setiferous anteriorly. Last pair of pereiopoda comparatively more elongated than in *C. curvispinum*, exceeding half the length of the body. Uropoda and telson nearly as in that species. Length of adult female 7 mm., of male 8 mm.

Remarks. — This species is nearly allied to the preceding one, though easily distinguishable by the more robust form of the body and by the structure of the 2 pairs of antennæ, the inferior of which are in both sexes very coarsely built, and have the last peduncular joint, as in *C. chelicorne*, produced to a spiniform process.

Description. — The length of adult, ovigerous females is about 7 mm., that of males 8 mm., and this form is accordingly somewhat larger than *C. curvispinum*, or about the size of *C. chelicorne*.

The form of the body (see fig. 10) is rather robust, with broad, flattened back.

The cephalon has the frontal edge but very slightly produced in the middle, forming an obtuse angle. The lateral lobes are moderately prominent and narrowly rounded at the tip.

The coxal and epimeral plates do not exhibit any difference from those in the 2 preceding spines.

The urosome (fig. 15) likewise agrees with that of the said species in having the last 2 segments less distinctly marked off from each other.

The eyes are small, but distinct, with dark pigment.

The superior antennæ are in the female comparatively short, not attaining $\frac{1}{3}$ of the length of the body, and have the 1st joint of the peduncle armed below with 3 distant spinules. In the male these antennæ (fig. 11) are, as usual, more fully developed, though not nearly so much elongated as in the male of *C. curvispinum*, and as in the latter, have the peduncle densely clothed with slender bristles. The 2nd joint is somewhat longer in the male than in the female, but in both sexes it is considerably shorter than the 1st. The flagellum in none of the sexes attains the length of the peduncle, and is composed of about 12 articulations.

The inferior antennæ are less different in the two sexes than is the case in *C. curvispinum*, exhibiting in both of them a very robust structure. In the male, however, they are (see fig. 10), as usual, somewhat coarser than in the female, exceeding somewhat in length $\frac{2}{3}$ of the body. The penultimate joint of the peduncle is very large and tumid, nearly as long as the last joint and the flagellum combined, and is produced at the end posteriorly to a moderately long, and but slightly curved spiniform projection, at the base of which there is a small, slightly bilobed expansion, similar to that found in *C. curvispinum*. The last joint of the peduncle has somewhat above the middle posteriorly a short recurved prominence, and the end of the joint is produced to a strong spiniform process similar to that in *C. chelicorne*. The flagellum is comparatively short, scarcely equalling in length the last peduncular joint, and exhibits the usual structure.

The gnathopoda do not exhibit any peculiarity in their structure.

The 2 anterior pairs of pereiopoda (fig. 13) on the whole resemble those in *C. curvispinum*, though they are somewhat more elongated. The basal joint is pronouncedly laminar and edged anteriorly with long setæ. The meral joint is about the length of the last 2 combined and rather broad, being in the male densely clothed with bristles anteriorly.

The last pair of pereiopoda (fig. 14) appear somewhat more elongated than in *C. curvispinum*, considerably exceeding half the length of the body, but otherwise they exhibit a very similar structure.

Also the uropoda and the telson are but little different, though, on a closer comparison, small differences may be stated to exist. Thus in comparing the 2nd pair of uropoda (fig. 16) with those in *C. curvispinum* (fig. 9), the rami are found to be comparatively longer and also armed with a greater number of spines.

The pigmentation of the body is very distinct and of a darker hue than in the other species.

Occurrence. — Of this species only a few specimens were collected by Mr. Warpachowsky at Stat. 32, about midway between the peninsula Mangyschlak and the opposite western coast. Some other specimens were collected last summer at Stat. 83, probably located in the eastern part of the North Caspian Sea.

In the collection of Dr. Grimm the species is represented by rather numerous specimens, partly collected in the Bays of Baku and Schachowaja from shallow water, partly in the middle part of the Caspian Sea from depths ranging from 7 to 40 fathoms.

27. Corophium mucronatum, G. O. Sars, n. sp.

(Pl. XXIV, figs. 1—7).

Specific Characters. — Frontal edge of cephalon angularly produced in the middle, lateral lobes comparatively short. Superior antennae of moderate length, and in both sexes but sparingly setous, 1st joint of the peduncle exceeding the other 2 combined, and having below 3 distant spinules, flagellum equalling in length the peduncle. Inferior antennae in male rather strong, with the penultimate joint considerably tumefied, subfusiform, and produced at the end posteriorly to a long mucroniform projection reaching beyond the midle of the last joint, and having at the base a small secondary tooth, last joint somewhat shorter than the penultimate one, and provided near the base posteriorly with a short recurved prominence, but without any spiniform process at the end. Anterior pairs of perciopoda moderately strong, meral joint rather much expanded distally, and clothed anteriorly with slender bristles. Last pair of perciopoda with the outer joints unusually broad, sublaminar. Uropoda and telson of the usual structure. Length of adult male 6 mm.

Remarks. — At first sight the present species somewhat resembles *C. chelicorne*, but is, on closer examination, easily distinguished by the very slender mucroniform projection of the penultimate peduncular joint of the inferior antennae, and by the want of a spiniform process at the end of the last peduncular joint. Moreover, this species is very prominently distinguished by the structure of the last pair of perciopoda, the outer joints of which exhibit a quite unusual broad, sublamellar shape.

Description of the adult male. — The length of the body in an apparently full-grown specimen scarcely attains 6 mm., and this form is accordingly somewhat inferior in size to the preceding ones.

The form of the body (see fig. 1) is that characteristic of the genus, being on the whole not very slender.

The cephalon has the frontal edge (see fig. 2) considerably produced between the bases of the superior antennae, forming in the middle an acute angle. The lateral lobes are comparatively short and narrowly rounded at the tip.

The coxal and epimeral plates exhibit the usual appearance.

The urosome (fig. 7), as in the 3 preceding species, has the line of demarcation between the last 2 segments less distinct than that between the 1st and 2nd.

The eyes are small, rounded, with dark pigment.

The superior antennæ (fig. 3) are rather elongated, considerably exceeding in length ⅓ of the body, and are but sparingly setiferous. The 1st joint of the peduncle is a little longer than the other 2 combined, and has below 3 distant spinules. The flagellum about equals in length the peduncle, and is composed of 12 articulations.

The inferior antennæ (see figs. 1 and 4) are rather strongly built, though scarcely exceeding half the length of the body. The penultimate joint of the peduncle is considerably tumefied, almost fusiform in shape, and is produced at the end posteriorly to a very long and slender, mucroniform projection extending beyond the middle of the last joint, and having at the base a small secondary tooth. The last peduncular joint is somewhat shorter than the penultimate one, and, as usual, much narrower, being cylindric in form. It is armed near the base posteriorly with a short recurved prominence, but has not any spiniform process at the end. The flagellum is a little shorter than the last peduncular joint, and of the usual structure.

The gnathopoda do not exhibit any peculiarity whatever.

The 2 anterior pairs of perciopoda (fig. 5) are moderately strong, with the basal joint pronouncedly laminar, and the meral joint considerably expanded distally, its anterior edge being clothed with scattered slender bristles. The 2 outer joints are not very slender, and the dactylus is about the length of the propodal joint.

The last pair of perciopoda (fig. 6) are about half the length of the body, and are prominently distinguished by the unusual shape of the outer joints, which, instead of being linear, are rather broad and compressed, and edged with fascicles of delicate bristles.

The uropoda and telson (see fig. 7) do not differ much from those parts in the other species.

The pigmentation of the body is the usual one, though it is less conspicuous than in *C. robustum*.

Occurrence. — Of this species some specimens, chiefly of the male sex, were collected by Mr. Warpachowsky at Stat. 63, in the eastern part of the North Caspian Sea. Solitary specimens were, moreover, taken at 2 other Stations (St. 53 and 56) north and west of the island of Kulaly.

In the collection of Dr. Grimm there are a few badly preserved specimens, collected partly in the Bay of Baku, partly in the bay of Balchansky from comparatively shallow water.

28. Corophium monodon, G. O. Sars, n. sp.

(Pl. XXIV, figs. 8—16).

Specific Characters. — Body rather slender, especially in the male. Frontal edge of cephalon angularly produced in the middle, lateral lobes narrowly rounded. Urosome with all the segments well defined. Superior antennæ of moderate length, and not very different in the two sexes, though the peduncle in male appears somewhat more elongated and more densely setous, 1st joint of the peduncle in both sexes longer than the other 2 combined, and having at the end below a single spinule; flagellum in female about the length of the peduncle, in male somewhat shorter. Inferior antennæ in female rather small and feeble, scarcely longer than the superior ones, in male much more elongated, exceeding $\frac{2}{3}$ of the length of the body, penultimate joint of the peduncle long and slender, almost cylindric in form, being produced at the end posteriorly to a narrow mucroniform projection not extending to the middle of the last joint, and having no secondary tooth at the base; last peduncular joint with only a very slight rudiment of a tooth near the base posteriorly; flagellum very short, scarcely exceeding half the length of the former joint. Anterior pairs of pereiopoda somewhat more slender than in *C. mucronatum*, meral joint gradually widening distally, and provided anteriorly with scattered bristles. Last pair of pereiopoda with the basal joint rather expanded, the outer joints, however, narrow and slender. Uropoda and telson of the usual structure. Length of adult female 4 mm., of male 5 mm.

Remarks. — Of all the Caspian species, this one would seem to come nearest to the typical species, *C. grossipes*, Lin. It is however evidently specifically distinct, differing, among other characters, in the much less strong development of the inferior antennæ, the penultimate peduncular joint of which is far less tumefied, and wants the deep sinus occurring in that species at the base of its terminal projection.

Description. — The length of fully adult, ovigerous females does not exceed 4 mm., that of the male being 5 mm., and this form accordingly is the smallest of the Caspian species, and in this respect is also rather inferior to the typical form, *C. grossipes*, Lin.

The form of the body (see fig. 8) is rather slender, especially in the male, otherwise of the usual appearance.

The cephalon has the frontal edge (see fig. 9) distinctly produced in the middle, forming an almost right angle. The lateral lobes are not very prominent and they are, as in the other Caspian species, narrowly rounded at the tip.

The coxal and epimeral plates are of the usual shape.

The urosome (fig. 15) has all the segments very distinctly defined, the line of demarcation between the last 2 segments being fully as sharply marked as that between the 1st and 2nd.

The eyes are comparatively larger than in the other species, and of a rounded form, with the pigment very dark.

The superior antennæ are in the female (see fig. 10) about $\frac{1}{2}$ of the length of the body, in the male, as usual, somewhat more elongated, though not nearly reaching half the length of the body. The peduncle is in the female but sparingly setous, whereas in the male (see figs. 8, 11) it is densely clothed below with slender bristles. In both sexes the 1st joint of the peduncle is considerably longer than the other 2 combined, and is armed below with a single spinule placed at the end of the joint. As in most other species, the 2nd peduncular joint is more elongated in the male than in the female (comp. fig. 10 and 11). The flagellum in the female about equals the peduncle in length, whereas in the male it is somewhat shorter. It is composed of from 10 to 12 articulations.

The inferior antennæ are in the female (see fig. 10) comparatively small and feeble, not even exceeding the superior ones in length. In the male (figs. 8 and 12) they are much more fully developed and rather slender, equalling about $\frac{2}{3}$ of the length of the body. The penultimate joint is scarcely at all dilated, being almost cylindric in form, and in the male nearly attains the length of the last peduncular joint and the flagellum combined. It is produced at the end posteriorly to a simple narrowly mucroniform projection, which does not nearly extend to the middle of the succeeding joint, and wholly wants any secondary tooth at the base. The last peduncular joint is in the female quite unarmed, whereas in the male there is a very slight rudiment of a dentiform prominence near the base posteriorly. The flagellum is comparatively very short, being in the male scarcely half as long as the last peduncular joint.

The gnathopoda exhibit the structure characteristic of the genus.

The 2 anterior pairs of pereiopoda (fig. 13) are somewhat more slender than in the 3 preceding species, though they resemble on the whole those in *C. mucronatum*. As in that species, the meral joint gradually widens distally, and is provided anteriorly with scattered slender setæ.

The last pair of pereiopoda (fig. 14) are moderately elongated, equalling about half the length of the body. The basal joint is rather large and expanded, whereas the outer joints exhibit the slender narrow form found in most other species.

The 2 anterior pairs of uropoda (see figs. 15), are constructed in the usual manner, though the difference in size is somewhat more pronounced in this than in most other species, the 2nd pair being very small as compared with the 1st.

The last pair of uropoda (fig. 16) have the terminal joint considerably narrower than the proximal one, and are only provided with a restricted number of bristles, between which a single apical spine is distinguished. In the typical species, *C. grossipes* Lin., this joint is much broader and lamelliform, without any spine.

The pigmentation of the body would seem to differ somewhat in different specimens, being as a rule restricted to the cephalon and the 6 anterior segments of the mesosome only, whereas the posterior part of the body appears almost devoid of pigment. On the antennae the pigment has a similar arrangement as is found in most other species.

Occurrence. — Of this species numerous specimens were collected by Mr. Warpachowsky at Stat. 63, lying in the eastern part of the North Caspian Sea. Solitary specimens were moreover taken at Stat. 64, in the neighbourhood of the former and at Stat. 59, in the western part of the basin.

In the collection of Dr. Grimm this species is represented by a few, in most cases very badly preserved specimens, which, according to the labels, were collected partly in the South Caspian Sea, partly in the Bays of Murawjew and Krasnowodsk, the greatest depth being 40 fathoms.

Explanation of the Plates.

Pl. XVII.
Niphargoides corpulentus, G. O. Sars.
(Figs. 1—13).

Fig. 1. Adult male, viewed from left side.
» 2. Superior antenna.
» 3. Inferior antenna.
» 4. Anterior gnathopod, with coxal plate.
» 5. Posterior gnathopod (basal joint not fully drawn).
» 6. First pereiopod.
Fig. 7. Antepenultimate pereiopod.
» 8. Penultimate pereiopod.
» 9. Last pereiopod.
» 10. Last epimeral plate, from left side.
» 11. Second uropod.
» 12. Last uropod.
» 13. Telson.

Niphargoides compactus, G. O. Sars.
(Figs. 14—19).

Fig. 14. Adult male, viewed from right side.
» 15. Superior antenna.
» 16. Outer part of an inferior antenna.
» 17. Anterior gnathopod (basal joint not fully drawn).
Fig. 18. Posterior gnathopod (do).
» 19. Last segment of urosome, with telson and right last uropod, dorsal view.

Pl. XVIII.
Niphargoides quadrimanus, G. O. Sars.
(Figs. 1—13).

Fig. 1. Adult, ovigerous female, viewed from left side.
» 2. Superior antenna.
» 3. Inferior antenna.
» 4. Anterior gnathopod, with coxal plate.
» 5. Posterior gnathopod.
» 6. Second pereiopod with coxal plate.
Fig. 7. Antepenultimate pereiopod.
» 8. Penultimate pereiopod.
» 9. Last pereiopod.
» 10. First uropod.
» 11. Second uropod.
» 12. Last uropod.
» 13. Telson.

Niphargoides æquimanus, G. O. Sars.
(Figs. 14—23).

Fig. 14. Adult male, viewed from right side.
» 15. Superior antenna.
» 16. Inferior antenna.
» 17. Anterior gnathopod, with coxal plate.
» 18. Posterior gnathopod (do).
» 19. Antepenultimate pereiopod (outer part not drawn).
Fig. 20. Last pereiopod.
» 21. Second uropod.
» 22. Last uropod.
» 23. Telson.

Pl. XIX.
Pandorites podoceroides, Grimm.

Fig. 1. Adult, ovigerous female, viewed from left side.
» 2. Cephalon, without the appendages.
» 3. Superior antenna.
» 4. Inferior antenna.
» 5. Anterior lip.
» 6. Posterior lip.
» 7. Left mandible, with palp.
» 8. Masticatory parts of the mandibles.
» 9. First maxilla.
» 10. Second maxilla.
» 11. Maxillipeds.
» 12. Anterior gnathopod, with coxal plate.
» 13. Posterior gnathopod, with coxal plate, branchial and incubatory lamellæ.
Fig. 14. Outer part of propodos of same, more highly magnified.
» 15. First pereiopod, with coxal plate.
» 16. Second pereiopod (do).
» 17. Antepenultimate pereiopod.
» 18. Penultimate pereiopod.
» 19. Last pereiopod.
» 20. First uropod.
» 21. Second uropod.
» 22. Last uropod.
» 23. Telson.
» 24. Adult male specimen (from Dr. Grimm's collection), viewed from right side.

Pl. XX.

Corophium nobile, G. O. Sars.

Fig. 1. Adult, ovigerous female, viewed from left side.
» 2. Adult male, from right side.
» 3. Superior antenna of female.
» 4. Inferior antenna of same.
» 5. Anterior lip.
» 6. Posterior lip.
» 7. Left mandible, with palp.
» 8. Right mandible, without the palp.
» 9. First maxilla.
» 10. Second maxilla.
Fig. 11. Maxillipeds.
» 12. Anterior gnathopod of female, with coxal plate.
» 13. Posterior gnathopod of same, with incubatory lamella.
» 14. Pleopod.
» 15. Inner corner of the basal part of same, more highly magnified, showing the peculiar structure of the 2 marginal spines.

Pl. XXI.

Corophium nobile, G. O. Sars

(continued).

Fig. 1. Adult male, viewed from the dorsal face.
» 2. Superior antenna of same.
» 3. Inferior antennæ of same.
» 4. Outer part of the flagellum, highly magnified, showing the 2 terminal hooks.
» 5. First pereiopod of male, with coxal plate and branchial lamella.
» 6. Antepenultimate pereiopod (do).
Fig. 7. Penultimate pereiopod.
» 8. Last pereiopod.
» 9. Urosome, without the left 1st and 2nd uropod; dorsal view.
» 10. First uropod.
» 11. Second uropod.
» 12. Last uropod.
» 13. Telson, from the dorsal face.
» 14. Same, viewed obliquely from right side, showing the vertical, dentated crests.

Pl. XXII.

Corophium chelicorne, G. O. Sars.

Fig. 1. Adult, ovigerous female, viewed from left side.
» 2. Frontal part of cephalon; dorsal view.
» 3. Superior antenna of same.
» 4. Inferior antenna of same.
» 5. Adult male, viewed from right side.
» 6. Superior antenna.
» 7. Inferior antenna.
» 8. Anterior gnathopod, with coxal plate.
Fig. 9. Posterior gnathopod.
» 10. First pereiopod.
» 11. Antepenultimate pereiopod, with branchial lamella.
» 12. Penultimate pereiopod.
» 13. Last pereiopod.
» 14. Urosome, without the right 1st and 2nd uropod; dorsal view.

Pl. XXIII.

Corophium curvispinum, G. O. Sars.

(Figs. 1—9).

Fig. 1. Adult male, viewed from right side.
» 2. Superior antenna of female.
» 3. Inferior antenna of same.
» 4. Anterior antenna of male.
» 5. Middle part of inferior antenna of same.
Fig. 6. First pereiopod.
» 7. Last pereiopod.
» 8. Urosome, without the right 1st and 2nd uropod; dorsal view.
» 9. Second uropod.

Corophium robustum, G. O. Sars.

(Figs. 10—16).

Fig. 10. Adult male, viewed from left side.
» 11. Anterior antenna of same.
» 12. Middle part of inferior antenna of same.
» 13. First pereiopod, with coxal plate and branchial lamella.
Fig. 14. Last pereiopod.
» 15. Urosome, without the left 1st and 2nd uropod.
» 16. Second uropod.

Pl. XXIV.

Corophium mucronatum, G. O. Sars.

(Figs. 1—7).

Fig. 1. Adult male, viewed from left side.
" 2. Frontal part of cephalon; dorsal view.
" 3. Anterior antenna of male.
" 4. Inferior antenna (basal part not fully drawn).
Fig. 5. First pereiopod, with coxal plate and branchial lamella.
" 6. Last pereiopod.
" 7. Urosome, without the left 1st and 2nd uropod; dorsal view.

Corophium monodon, G. O. Sars.

(Figs. 8—16).

Fig. 8. Adult male, viewed from right side.
" 9. Cephalon, without the appendages; dorsal view.
" 10. Cephalon of female, with antennæ and oral parts, viewed from right side.
" 11. Superior antenna of male.
Fig. 12. Inferior antenna of same.
" 13. First pereiopod, with coxal plate and branchial lamella.
" 14. Last pereiopod.
" 15. Urosome, viewed from the dorsal face.
" 16. Last uropod.

G.O.Sars Crustacea caspia.
Amphipoda. Pl. XVII.

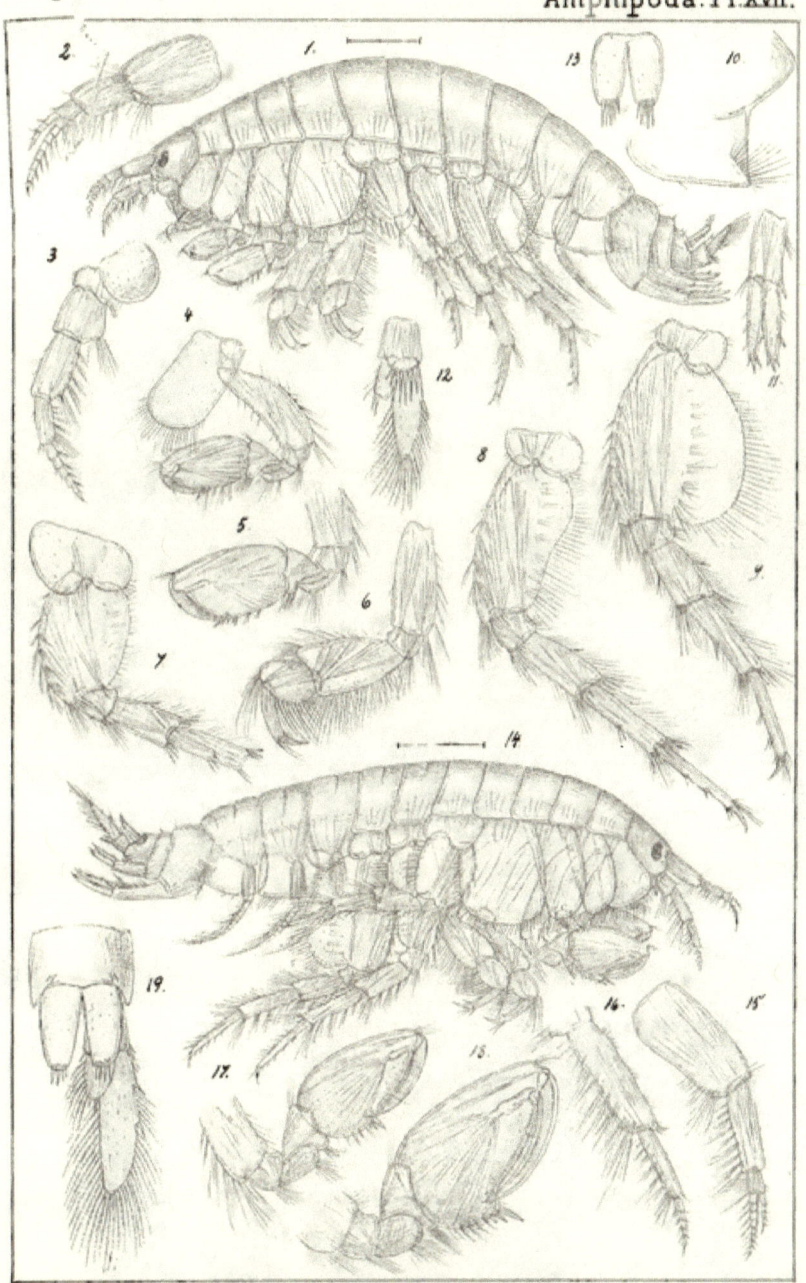

G.O Sars autogr.
Niphargoides corpulentus, n. sp.
Niphargoides compactus, n. sp.

G.O.Sars Crustacea caspia.
Amphipoda. Pl. XVIII

Niphargoides qvadrimanus, n. sp.
Niphargoides æqvimanus, n. sp.

G.O.Sars Crustacea caspia.
Amphipoda. Pl. XIX.

Pandorites podoceroides, Grimm.

G.O Sars Crustacea caspia.
Amphipoda. Pl XX.

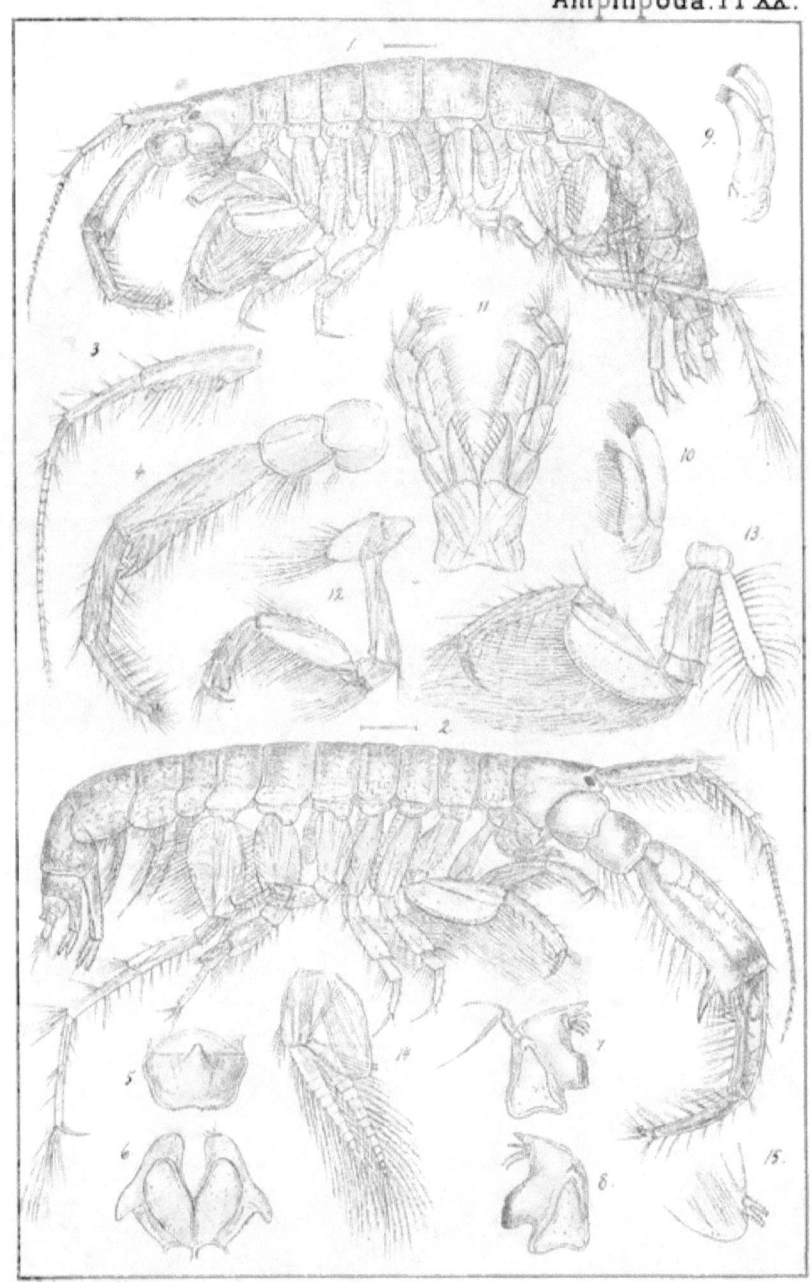

Corophium nobile, n. sp.

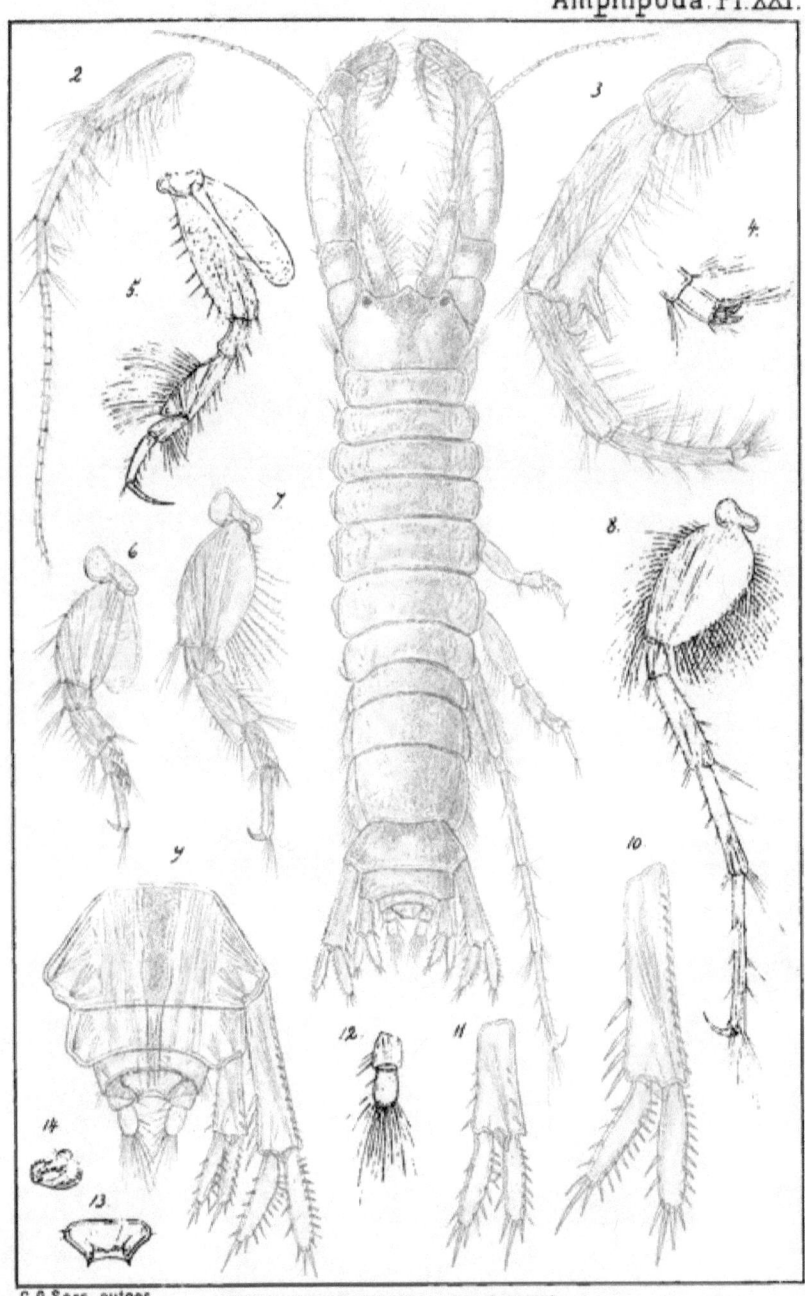

Corophium nobile, n. sp. (contin.).

G.O.Sars Crustacea caspia.
Amphipoda Pl. XXII.

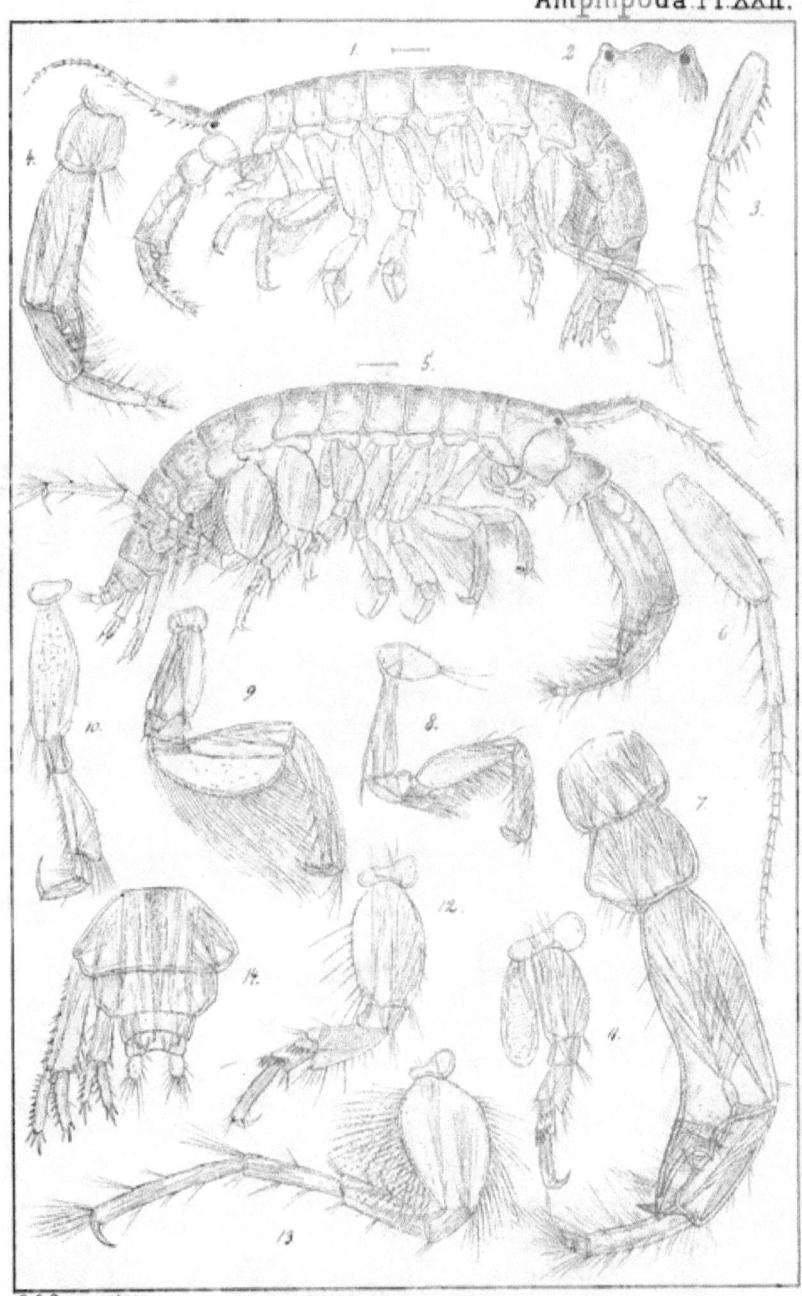

Corophium chelicorne, n. sp.

G.O.Sars Crustacea caspia.
Amphipoda. Pl. XXIII.

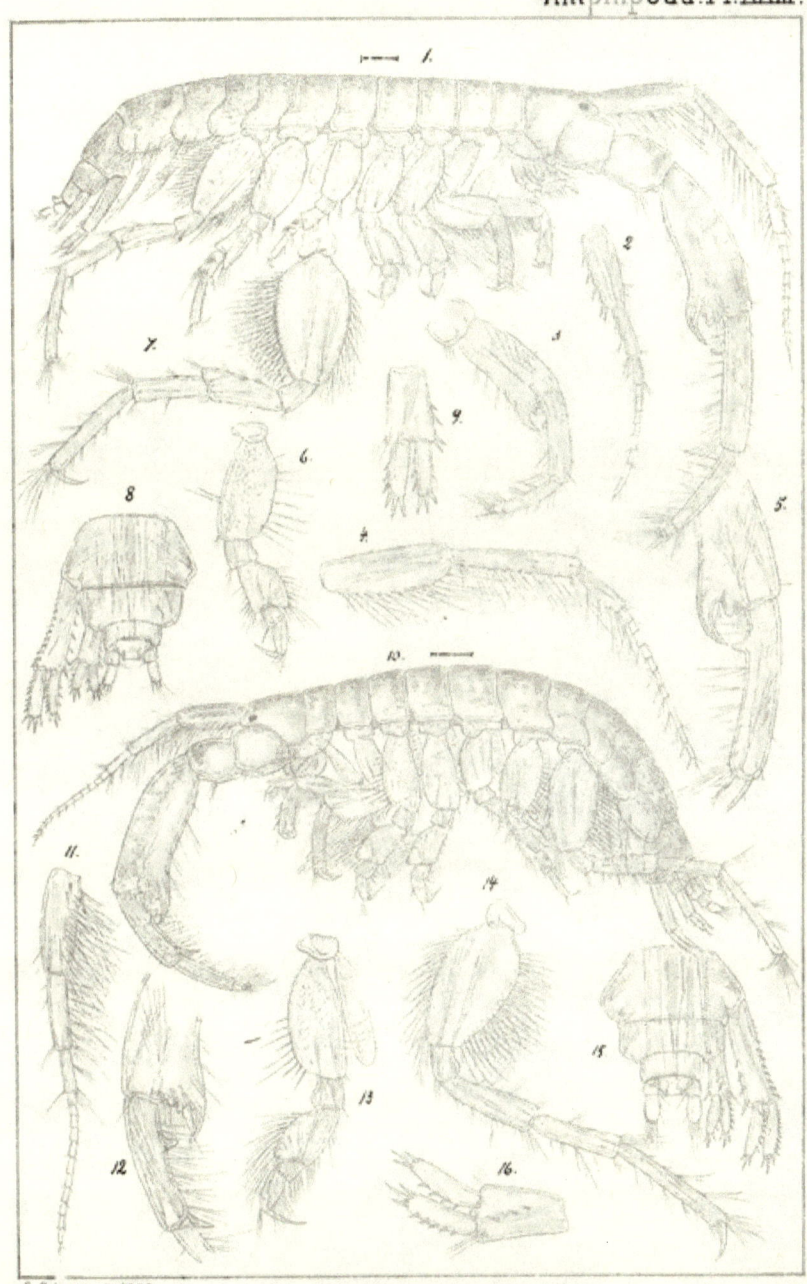

1. Corophium curvispinum, n.sp.
2. Corophium bidentatum, n.sp.

G.O.Sars Crustacea caspia.
Amphipoda. Pl. XXIV.

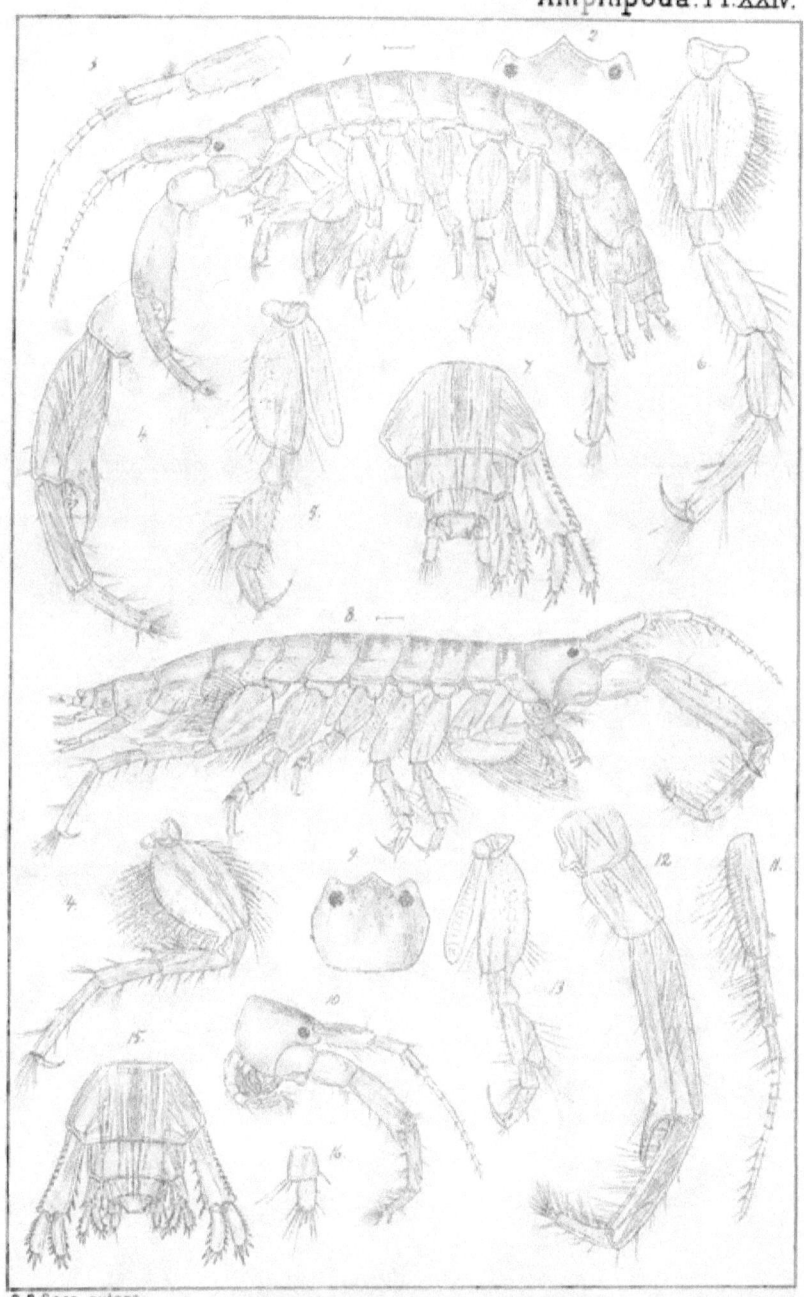

Corophium mucronatum, n. sp.
Corophium monodon, n. sp.

ОГЛАВЛЕНІЕ. — SOMMAIRE.

	Стр.		Pag.
Извлеченія изъ протоколовъ засѣданій Академіи	XXVII	Extraits des procès verbaux des séances de l'Académie	XXVII
Отчетъ о шестомъ присужденіи Академіею Наукъ премій митрополита Макарія за 1895 годъ	221	Compte rendu du VI concours des prix de l'archevêque-métropolitain Macaire	221
Ө. Бредихинъ. Вѣковыя возмущенія орбиты кометы 1862 III и ея производныхъ орбитъ	251	Th. Brédikhine. Variations séculaires de l'orbite de la comète 1862 III et de ses orbites dérivées	251
Д. Клеменцъ. Краткій отчетъ о путешествіи Д. Клеменца по Монголіи за 1894 годъ	261	D. Klementz. Compte rendu sommaire d'un voyage en Mongolie en 1894	261
Г. О. Сарсъ. Каспійскія ракообразныя. Матеріалы для изученія карцинологической фауны Каспійскаго моря. (Съ 8 таблицами рис.)	275	G. O. Sars. Crustacea caspia. Contributions to the knowledge of the Carcinological Fauna of the Caspian Sea. (With 8 autographic plates.)	275

Напечатано по распоряженію Императорской Академіи Наукъ.
Сентябрь 1895 г.   Непремѣнный секретарь, Академикъ *Н. Дубровинъ*.

ТИПОГРАФІЯ ИМПЕРАТОРСКОЙ АКАДЕМІИ НАУКЪ.
Вас. Остр., 9 линія, № 12.

ИЗВѢСТІЯ

ИМПЕРАТОРСКОЙ АКАДЕМІИ НАУКЪ.

ТОМЪ IV. № 5.

1896. МАЙ.

BULLETIN

DE

L'ACADÉMIE IMPÉRIALE DES SCIENCES

DE

ST.-PÉTERSBOURG.

Vᵉ SÉRIE. TOME IV. № 5.

1896. MAI.

С.-ПЕТЕРБУРГЪ. — St.-PÉTERSBOURG.
1896.

Crustacea Caspia.

Contributions to the Knowledge of the Carcinological Fauna of the Caspian Sea.

By

G. O. Sars,

Professor of Zoology at the University of Christiania, Norway.

AMPHIPODA.

Supplement.

With 12 autographic plates.

INTRODUCTION.

In the present paper, I propose to describe and figure a number of additional Caspian Amphipoda, the greater part of which are derived from the rich collection of Dr. Grimm, kindly placed in my hands for examination. To these are added some species collected by Mr. Warpachowsky in the northern part of the Caspian Sea, and a few others procured by Messrs. Maximovicz and Andrussow from the southern part of the basin. Some of the species are unfortunately only represented by solitary specimens, and the examination of these has of course not been so close as might have been desired. Five of the species established by Dr. Grimm are altogether omitted, because they are only represented by immature, or very badly preserved specimens, the examination of which has not led to any definite result. These species are named by Dr. Grimm as follows: *Gammarus priscus, G. portentosus, G. multiformis, Pandora cocca, Iphigencia abyssorum*. Three other species are named by the said author in a short treatise inserted in the Archiv für Naturgeschichte 1880, viz., *Gammarus Gregorkowii, G. coronifer* and *G. thaumops;* but none of these species are represented in his collection.

The number of additional species described in the present paper is 25. Together with those previously recorded by the present author, the total number of Caspian Amphipoda, which up to this time have been examined more closely, amounts to no less than 53, belonging to 4 different families, viz., *Lysianassidæ*, *Pontoporeiidæ*, *Gammaridæ* and *Corophiidæ*, the *Gammaridæ* being the family by far the most abundantly represented. At the close of this paper, I shall give a complete list of all the species.

Fam. LYSIANASSIDÆ.
Gen. **Pseudalibrotus**, Della Valle.

Syn.: *Alibrotus*, G. O. Sars (olim).

Remarks. — In his great work on the Gammarids of the gulf of Naples, Mr. Della Valle observes that the well-known arctic form *Anonyx littoralis* Kröyer, which I had referred, though with some hesitation, to the genus *Alibrotus* of M-Edwards, is scarcely congeneric with the type of the latter genus, *A. chausseica*. For this reason, he has established the new genus *Pseudalibrotus*, to receive the arctic species, and in this view I am now much inclined to agree with that distinguished author. The genus *Pseudalibrotus* thereby becomes an exclusively arctic one. Now it is rather interesting that this genus is also represented in the Caspian Sea by 2 apparently distinct species, to be described below.

1. **Pseudalibrotus caspius**, (Grimm).
(Pl. 1, figs. 1—20).

Onesimus caspius, Grimm MS.
" *pomposus*, Grimm δ.

Specific Characters. — Body rather stout and somewhat compressed, with the back strongly curved. Lateral lobes of cephalon narrowly rounded. First pair of coxal plates much broader than the succeeding ones, and somewhat expanded distally; 2nd and 3rd pairs very narrow; 4th pair but very little expanded; 5th pair nearly as broad as they are deep. Last pair of epimeral plates of metasome acutely produced at the lateral corners. First segment of urosome slightly depressed dorsally. Eyes comparatively small, rounded oval. Antennæ somewhat shorter than in the type species, the superior ones with the flagellum twice the length of the peduncle, accessory appendage 4-articulate. Anterior gnathopoda not very powerful, propodos scarcely as broad as the carpus, and somewhat attenuated distally, palm short, oblique, with a single slender spine at the inferior corner. Posterior gnathopoda with the propodos obliquely produced at the tip. Pereiopoda rather short and thick, and but sparsely setiferous; last pair with the basal joint about the length of the remaining part of the leg, oval, broadest in its upper part, infero-posteal corner produced to a rounded lobe, posterior edge divided into about 7 serrations. Last pair of uropoda with both rami simple mucronate. Telson with the tip slightly notched on each side and obtusely produced in the middle. Length 9 mm.

Remarks. — The present form, detected by Dr. Grimm, is unquestionably congeneric with the arctic species, *P. littoralis*, though easily distinguishable by the less elongated antennæ, and the somewhat different structure of the gnathopoda. Also the pereiopoda, uropoda and telson exhibit some points of difference. One specimen in the collection of Dr. Grimm, named *Onesimus pomposus*, I regard as only a young male of the present species.

Description. — The length of an apparently adult female specimen measures about 9 mm., and this form is accordingly somewhat inferior in size to the arctic species.

The general form of the body (see fig. 1) resembles that of the said species, though being perhaps somewhat shorter and stouter, with the back generally much curved and, as in most other Lysianassidæ, quite smooth.

The cephalon (fig. 4) somewhat exceeds in length the 1st segment of the mesosome, and has the lateral lobes slightly produced, and narrowly rounded at the tip. The postantennal corners are distinct and nearly rectangular, and behind them there is on each side an angular incision, into which the upper end of the mandible is received, the incision being defined behind by an acute lappet.

The anterior pairs of coxal plates are somewhat deeper than the corresponding segments, and densely crowded. The 1st pair (see fig. 11) are considerably broader than the others, and widen somewhat distally, with the end obtusely truncated and provided at the posterior corner with a small dentiform projection. The 2 succeeding pairs (see fig. 13) are remarkably narrow, the 4th pair (see fig. 14) also being of inconsiderable breadth as compared with most other Lysianassidæ.

The 3 posterior pairs of coxal plates exhibit the usual shape, the antepenultimate pair (see fig. 15) being rather broad, of a rounded quadrangular form.

The epimeral plates of the metasome are well developed, the 1st pair being, as usual, rounded, whereas the other 2, and especially the last one, are rather acutely produced at the lateral corners.

The urosome is quite smooth, and has the 1st segment very slightly depressed above, in front of the middle.

The eyes (see fig. 4) are comparatively small and of a rounded oval form. The pigment would seem to have been of the usual bright red colour, but in all the specimens it was quite absorbed by the action of the spirit.

The superior antennæ (fig. 2) about equal in length the cephalon and the first 3 segments of the mesosome combined, and are accordingly somewhat shorter than in the arctic species. The peduncle is short and thick, with the 1st joint very large, fully twice as long as the other 2 combined, and somewhat applanated. The flagellum is about twice as long as the peduncle, and is composed of 16 articulations, the 1st of which is much the largest. It is provided in its proximal part outside with bundles of delicate olfactory filaments, 5 of which belong to the 1st joint. The accessory appendage is about half the length of the peduncle, and composed of 4 articulations, the 1st of which nearly equals in length the other 3 combined.

The inferior antennæ (fig. 3) are somewhat longer than the superior, and have the basal joint rather swollen, and transversely elliptical in form. The penultimate joint of the peduncle is considerably larger than the last one, and, like the latter, provided anteriorly with fascicles of delicate bristles. The flagellum considerably exceeds in length that of the superior antennæ, and is composed of about 20 articulations.

The buccal mass (see fig. 4) is rather protruding, though to a great extent covered laterally by the 1st pair of coxal plates (see fig. 1). The oral parts on the whole resemble in structure those in the type species.

The anterior lip forms in front a compressed rounded lamella (fig. 5), which, however, is not very prominent.

The posterior lip (fig. 6) has the lateral lobes slightly bilobed at the tip, and exhibits on each side a narrowly rounded auricular expansion. The lobes are finely ciliated both at the tip and along the inner edge.

The mandibles (fig. 7) are rather strong, and have the cutting part simple, with only a slight trace of a denticle on each corner. The molar expansion is well defined, though not very large, and exhibits at the tip a distinct triturating surface. The palp is affixed at the same level as the molar expansion, and about equals in length the mandible. Its terminal joint is somewhat shorter than the 2nd, and oblong oval in form, having the usual supply of bristles. The 2nd joint, on the other hand, is almost naked, with only 2 small bristles near its end.

The 1st pair of maxillæ (fig. 8) almost exactly resemble those in the type species, having the masticatory lobe rather large and lamellar, with a dense assemblage of partly denticulated spines at the anterior corner, and behind them 3 smaller spines issuing from slight notches of the densely ciliated margin. The basal lobe is rather small and has 2 ciliated apical setæ. The palp is not very large, with the terminal joint but little expanded.

The 2nd pair of maxillæ (fig. 9) have the outer lobe considerably larger than the inner, and only setous at the tip. The inner lobe is finely ciliated inside, and carries, in addition to the terminal bundle of bristles, a strong spine somewhat beyond the middle.

The maxillipeds (fig. 10) have the palps less strongly developed than in the arctic species, but otherwise exhibit a very similar structure.

The anterior gnathopoda (fig. 11) are not nearly so strong as in *P. littoralis*, differing also conspicuously in the shape of the propodos. The latter (see also fig. 11a) is scarcely as broad as the carpus, and is not at all expanded, being on the contrary somewhat narrowed distally. The palm is rather short and somewhat oblique, with only a single slender spine issuing from the lower corner. The dactylus is also much smaller than in the said species.

The posterior gnathopoda (fig. 12) are very slender, and likewise differ somewhat from those in *P. littoralis* in the shape of the propodos (fig. 12a), which is not, as in that species, transversely truncated at the tip, but is somewhat produced at the lower corner, where the minute chela is formed.

The pereiopoda are considerably shorter and also less densely setous than in the arctic species. The 2 anterior pairs (fig. 13) are more slender than the posterior ones, and have the meral joint slightly produced at the end anteriorly. The propodal joint is about the length of the 2 preceding joints combined. The antepenultimate pair (fig. 15) have the basal joint irregularly oval, and like that of the 2 posterior pairs, produced at the infero-posteal corner to a rounded lobe. Along its anterior edge is a row of about 10 short spines. The 3 succeeding joints are comparatively short and, combined, but little longer than the propodal joint. The penultimate pair (fig. 16) somewhat exceed in length the antepenultimate one, and have the basal joint comparatively more elongated, though scarcely broader.

The last pair (fig. 17) are a little shorter than the penultimate one, and have the basal joint rather large, fully as long as the remaining part of the leg, and of an oval form, slightly narrowed distally. The posterior edge of the joint, as in the 2 preceding pairs, is divided into a limited number of serrations, scarcely more than 7.

The 2 anterior pairs of uropoda (fig. 18) are of similar structure, though somewhat differing in size, the rami in both pairs being simple mucroniform, and provided with scattered spines.

The last pair of uropoda (fig. 19) project but slightly beyond the others, and have the basal part comparatively short, otherwise agreeing with the preceding pairs in the simple mucronate shape of the rami.

The telson (fig. 20) is entire, squamiform, about as long as it is broad at the base, and has the tip slightly produced in the middle with a small notch on each side, from which a minute hair issues. In the arctic species the tip exhibits traces of a slight emargination.

Occurrence. — This form has been collected by Dr. Grimm in 4 Stations belonging to the middle part of the Caspian Sea, the depth ranging from 80 to 250 fathoms.

2. Pseudalibrotus platyceras (Grimm).

(Pl. 1, figs. 21—23).

Onesimus platyceras, Grimm MS.

Specific Characters. — Very like the preceding species, but about twice as large and comparatively more tumid, with the back broadly rounded. Lateral lobes of cephalon subangular. Anterior pairs of coxal plates comparatively broader than in *P. caspius*. Last pair of epimeral plates of metasome less acutely produced. Urosome with a conspicuous saddle-like depression across the anterior part of the 1st segment. Eyes narrowed in their upper part. Superior antennæ with the 1st joint of the peduncle very large and applanated, flagellum composed of a greater number of articulations than in the preceding species, accessory appendage exceeding half the length of the peduncle, and 6-articulate. Gnathopoda, pereiopoda, uropoda and telson apparently of a similar structure to that in *P. caspius*. Length nearly 20 mm.

Remarks. — This form is very nearly allied to the preceding one, but apparently distinct, being fully twice as large. For want of specimens, however, a closer anatomical comparison of the two could not be instituted.

Description. — The length of an adult female specimen attains to nearly 20 mm., and this form accordingly grows to a much larger size than the preceding one, and in this respect even exceeds the arctic species.

The form of the body (see fig. 21) on the whole closely resembles that of the preceding species, though still more robust and rather tumid, with the back broadly rounded.

The cephalon but slightly exceeds in length the 1st segment of the mesosome, and has the lateral lobes subangular in front.

The coxal plates appear somewhat broader than in the preceding species, though exhibiting an exactly similar mutual relation.

The last pair of epimeral plates of the metasome have the lateral corners pointed, but comparatively less produced than in *P. caspius*. The urosome has across its 1st segment, somewhat in front of the middle, a very distinct saddle-like depression.

The eyes are somewhat less regularly oval than in the preceding species, having their upper part narrowed to an obtuse point.

The superior antennæ are about the length of the cephalon and the 4 anterior segments of the mesosome combined, and have the 1st joint of the peduncle (see fig. 22) exceedingly large, and conspicuously applanated. The flagellum is rather slender and composed of a much greater number of articulations than in the preceding species, amounting to from 30 to 40. The accessory appendage is likewise more fully developed, consisting of 6 articulations.

The inferior antennæ are but little longer than the superior, and have the flagellum divided into 34 articulations.

The anterior gnathopoda are constructed as in the preceding species, but have the propodos (see fig. 23) somewhat shorter in proportion to its breadth and less obliquely truncated at the tip.

The posterior gnathopoda could not be examined more closely, as they were hidden between the coxal plates.

The pereiopoda seem on the whole to agree very closely with those in *P. caspius*, but are more densely setiferous and have the posterior edge of the basal joint in the 3 posterior pairs divided into a greater number of serrations.

As to the uropoda and telson, they do not seem to differ in any essential manner from those parts in *P. caspius*.

Occurrence. — Of this form 2 specimens are contained in the collection of Dr. Grimm, the one preserved in spirit, the other in glycerine, both being in a very good state of preservation, which prevented me from sacrificing either of them for a closer anatomical examination. The specimens, which exactly agree both as to size and other characters, were collected by Dr. Grimm at Stat. 124, belonging to the middle part of the Caspian Sea, from a depth of 40 fathoms.

Fam. PONTOPOREIIDÆ.

Gen. **Pontoporeia**, Kröyer.

Remarks. — Of this genus only 2 species have hitherto been recorded, the one, *P. femorata*, Kröyer, being an exclusively marine form, whereas the other, *P. affinis* Lindström, also occurs in the great lakes of Norway,

Sweden and Russia, as also of North America. Both species are certainly of true arctic origin. To these species may now be added a 3rd, peculiar to the Caspian Sea, and well defined from any of the others.

3. Pontoporeia microphthalma, Grimm. MS.

(Pl. 2, figs. 1—7).

Specific Characters. — Body moderately slender and somewhat compressed. Cephalon with the lateral lobes narrowly rounded. Coxal and epimeral plates nearly as in *P. affinis*. Urosome rather short and stout, with the 1st segment considerably elevated dorsally at the posterior edge, and having in the middle 2 minute juxtaposed denticles, on each side a single small spine. Eyes comparatively small, irregularly oval in form. Antennæ nearly as in *P. affinis*. Anterior gnathopoda with the carpus moderately expanded, propodos oblong oval, with the palm rather short and imperfectly defined below. Posterior gnathopoda with the propodos narrow oblong and transversely truncated at the tip. Last pair of pereiopoda with the basal joint very large, rounded oval, posterior edge regularly curved throughout. Uropoda and telson of the usual structure. Length 6 mm.

Remarks. — The present new species, detected by Dr. Grimm, is nearly allied to *P. affinis* Lindström, but easily distinguishable by the comparatively smaller eyes, the dorsally produced 1st segment of the urosome, and the more regular form of the basal joint of the last pair of pereiopoda. Also in the form of the propodos of the 2 pairs of gnathopoda, some differences are to be found.

Description. — The length of the only tolerably well preserved specimen, apparently an adult female, scarcely exceeds 6 mm., and this form is accordingly rather inferior in size to the other 2 known species.

The general form of the body (see fig. 1) resembles that in the other 2 species, being moderately slender, and somewhat compressed, with the mesosome and metasome quite smooth throughout.

The cephalon about equals in length the first 2 segments of the mesosome combined, and is somewhat produced in front between the bases of the superior antennæ (see fig. 3). The lateral lobes are slightly prominent, and narrowly rounded at the tip.

The anterior pairs of coxal plates are somewhat deeper than the corresponding segments, and densely fringed at the end with ciliated setæ. The 1st pair are a little broader than the 2nd, but scarcely at all expanded distally. The 4th pair are slightly emarginated posteriorly, and somewhat expanded in their outer part, forming beneath the emargination an obtuse-angular corner.

The posterior pairs of coxal plates are much less deep than the anterior, and successively diminish in size. The antepenultimate pair have the anterior lobe somewhat deeper than the posterior, and evenly rounded at the tip.

The epimeral plates of the metasome are well-developed, the 1st pair being, as usual, rounded, the 2 posterior pairs nearly rectangular. The urosome is rather short and stout, and highly remarkable by the great elevation of the posterior edge of the 1st segment above the level of the 2nd. On the most prominent place 2 small juxtaposed denticles occur, and beside these, there is, on the same segment, a small lateral spine. In *P. affinis*, this segment, like the others, is quite smooth, and in *P. femorata* it gives rise to the peculiar forked dorsal process, characteristic of that species.

The eyes are (see fig. 3) very small, and of a somewhat irregular oval form, with the pigment rather light.

The antennæ resemble in structure those in the 2 other species, being rather strongly built, and subequal in length, with a number of finely ciliated setæ issuing from the hind edge of the outer peduncular joints. The superior ones about equal in length the cephalon and the 3 first segments of the mesosome combined, and have the 1st joint of the peduncle very large, exceeding in length the other 2 combined. The flagellum (see fig. 2) about equals in length the peduncle, and is composed of 9 articulations, the 1st of which is much the largest. The accessory appendage somewhat exceeds in length the last peduncular joint, and is composed of 3 well-defined articulations.

The inferior antennæ are somewhat more strongly built than the superior, almost pediform, and have the 3 outer joints of the peduncle rather thick, with densely crowded bristles both anteriorly and posteriorly. The flagellum is about the same length as that of the superior antennæ, and is composed of a similar number of articulations.

The anterior gnathopoda (fig. 5) have the carpus rather broad and expanded, though not nearly so much as in *P. femorata*. The propodos is considerably narrower than the carpus, and oblong oval in form, with the palm quite short, and imperfectly defined below. In the other 2 species the palm is much more elongated, occupying the greater part of the inferior edge. The dactylus is comparatively very small.

The posterior gnathopoda (fig. 6) are more slender and elongated than the anterior, with the propodos oblong linear in form, and having the palm transverse, the lower corner being rectangular, not, as in the other 2 species, produced to a thumb-like prominence.

The 2 anterior pairs of perciopoda are of quite normal structure.

The 3 posterior pairs exhibit the structure characteristic of the genus, and are rather unequally developed, the penultimate pair being much more elongated than the other 2, and generally strongly reflexed. The basal joint on this and the preceding pair is rather narrow, and somewhat tapering distally; whereas on the last pair (fig. 7) it is extremely large and expanded, being fully as long as the remainder of the leg. The shape of the joint somewhat differs from that in the other 2 species, being more regularly rounded oval, with the posterior edge quite evenly curved and densely fringed with ciliated setæ.

The uropoda and telson do not seem to differ materially from those parts in the 2 other species.

Occurrence. — Of this species 2 specimens, the one imperfect, are contained in the collection of Dr. Grimm, having been taken, according to the label, at St. 108, belonging to the middle part of the Caspian Sea, the depth ranging from 80 to 90 fathoms.

Fam. GAMMARIDÆ.

Gen. **Gmelina**, Grimm.

Of this genus, established by Dr. Grimm, 2 species have been described by the present author in his 1st article on the Caspian Amphipoda. Two additional species are now added, and will be described below, both of them easily distinguishable from the 2 first recorded.

4. **Gmelina laeviuscula**, G. O. Sars, n. sp.

(Pl. 2, figs. 8—12).

Specific Characters. — Body moderately slender and nearly smooth, without any distinct tubercles or dorsal expansions, though having the segments rather sharply marked off from each other. Cephalon with the lateral faces quite smooth, lateral lobes narrowly rounded in front. Anterior pairs of coxal plates (in male) but little deeper than the corresponding segments, and of the usual shape. Last pair of epimeral plates of metasome but slightly produced at the lateral corners. Urosome with a few small hairs above at the end of each segment. Eyes very small, rounded oval. Antennæ of the usual structure. Gnathopoda (in male) rather powerful. Pereiopoda nearly as in the 2 first described species. Last pair of uropoda very robust, though not attaining the length of the urosome, outer ramus unusually broad, foliaceous, and densely fringed with fascicles of slender spines, inner ramus less rudimentary than in the other 2 species. Telson comparatively large, cleft very deep and fissure-like, lateral lobes obtusely truncated at the tip and having several lateral bristles and a dense row of apical spines. Length of adult male 7 mm.

Remarks. — This new species is at once distinguished from the 2 previously described forms by the body having no distinct lateral tubercles or dorsal expansions. In the structure of the several appendages it nearly agrees with them, though the last pair of uropoda and the telson exhibit well-marked specific differences.

Description. — The solitary specimen examined, which is an adult male, measures in length about 7 mm., and this form is accordingly somewhat inferior in size to the 2 species previously described by the present author.

The body (see fig. 8) is moderately slender and, as in the other species, much compressed, without, however, exhibiting any distinct lateral tubercles. Nor are any of the segments elevated to dorsal expansions, though they appear rather sharply marked off from each other.

The cephalon does not attain the length of the first 2 segments of the mesosome combined, and is somewhat narrowed in front, with the lateral faces quite smooth. The rostral projection (see fig. 10) is well marked, and the lateral lobes slightly produced and narrowly rounded in front.

The anterior pairs of coxal plates are but little deeper than the corresponding segments, and of a shape similar to that in the 2 previously described species; the 1st pair not being at all expanded distally, and the 4th not much broader than the preceding ones. The 3 posterior pairs are rather small.

The epimeral plates of the metasome are normally developed, the last pair being only slightly produced at the lateral corners.

The urosome is of moderate size, and has dorsally at the end of the segments a few small hairs.

The eyes (see fig. 10) are very small, and placed at some distance from the lateral corners of the cephalon. They are oval in form and have a dark pigment.

The antennæ exhibit the structure characteristic of the genus, being rather slender and feeble, though not much elongated. The superior ones somewhat exceed in length the inferior, and have the peduncle somewhat longer than the flagellum, which is composed of about 10 articulations. The accessory appendage (see fig. 9), as in the other species, is extremely small, and uniarticulate. The inferior antennæ have the last peduncular joint somewhat shorter than the penultimate, and the flagellum about half the length of the peduncle.

The gnathopoda (see fig. 8) are rather powerfully developed, and show the specimen to be an adult male. In structure they seem to agree rather closely with those in the males of the other species.

The pereiopoda also do not seem to exhibit any marked difference from those in the 2 previously described species.

The 2 anterior pairs of uropoda are very unequal in size, the 1st pair being fully twice as long as the 2nd. Their structure is, however, the usual one.

The last pair of uropoda (fig. 11) are rather robust, and project considerably beyond the others, though not quite attaining the length of the urosome. The outer ramus is unusually broad, foliaceous, and exhibits a rather small terminal joint. On the outer edge of the ramus there are about 6 dense fascicles of slender spines, and on the inner edge 4 similar ones. The inner ramus is less rudimentary than in the other species, being almost half the length of the outer, and carries on the tip several slender spines.

The telson (fig. 12) is comparatively large, and oval in form, and is divided by a narrow, fissure-like cleft into two halves, each of which has from 4 to 5 slender lateral spines. The tip of the lateral lobes is obtusely truncated and armed with a dense row of about 7 slender spines.

Occurrence. — The above-described specimen was taken by Mr. Warpachowsky at Stat. 69, probably in the eastern part of the North Caspian Sea.

5. Gmelina pusilla, G. O. Sars, n. sp.

(Pl. 2, figs. 13—21).

Specific Characters. — ♀. Body rather short and stout, somewhat compressed, and perfectly smooth throughout. Cephalon with the lateral lobes angularly produced in front. Anterior pairs of coxal plates rather deep and fringed distally with scattered bristles; 1st pair obliquely expanded, so as to form a rounded lobe extending forwards; 4th pair somewhat less deep than the preceding ones, and scarcely broader. Last pair of epimeral plates of metasome nearly rectangular. Urosome without any spines or hairs dorsally. Eyes of moderate size and placed unusually far down, close to the lateral corners of cephalon. Antennæ comparatively short, subequal in length. Gnathopoda rather feeble, propodos of the anterior ones oval, that of the posterior oblong quadrangular. Posterior pairs of pereiopoda comparatively short, basal joint of last pair rather large, irregularly oval, and densely setous both anteriorly and posteriorly. Last pair of uropoda not very large, outer ramus gradually tapering distally, inner ramus very small.

Telson of moderate size, deeply cleft, lateral lobes obtusely pointed, and having each 2 lateral, and 2 apical spines. Length of adult female 5 mm.

Remarks. — The external appearance of this form looks rather different from that in the other species, and I was therefore at first in some doubt about its true place. Having, however, subsequently examined some of the appendages more closely, I find it to be referable to the genus *Gmelina*, though representing a somewhat anomalous species.

Description. — The length of the solitary specimen examined, which is an adult female with fully developed incubatory lamellæ, measures only 5 mm., and this form is accordingly much the smallest of the 4 species as yet known.

The body (see fig. 13) is rather short and stout, somewhat compressed, and perfectly smooth throughout, without any trace of tubercles or dorsal expansions.

The cephalon (see fig. 14) nearly equals in length the first 2 segments of the mesosome combined, and is gradually narrowed in front. The rostral projection is rather small, though distinct, and the lateral lobes are angularly produced in front.

The anterior pairs of coxal plates are much deeper than the corresponding segments and edged with scattered bristles. The 1st pair (see fig. 16) are of a somewhat unusual form, being obliquely expanded, so as to form in front a linguiform lobe advancing over the oral parts and the basal joint of the inferior antennæ. The 2 succeeding pairs (see fig. 17) are oblong quadrangular in shape, and of nearly equal size. The 4th pair (see fig. 13) are less deep than the preceding ones, and scarcely broader, being obliquely truncated at the tip, with a very slight emargination behind. The posterior pairs are very small.

The epimeral plates of the metasome are normally developed, the 1st being rounded, the other 2 nearly rectangular.

The urosome is quite smooth, with no trace of spines or bristles dorsally.

The eyes (see fig. 14) are of moderate size, and rounded oval in form, with dark pigment and well-developed visual elements. They are placed unusually far down, close to the lateral corners of the cephalon.

The superior antennæ (fig. 15) are rather feeble, scarcely exceeding in length the cephalon and the first 2 segments of the mesosome combined, and have the 2nd joint of the peduncle about as long as the 1st, whereas the 3rd joint is scarcely half as long. The flagellum equals in length the peduncle, and is composed of 7 articulations. The accessory appendage is extremely small, and uniarticulate.

The inferior antennæ (see fig. 13) are somewhat more strongly built than the superior, but scarcely longer, and have the flagellum comparatively small, and 4-articulate.

The gnathopoda are rather feeble in structure, and densely supplied with bristles, the posterior ones (fig. 17) being a little more elongated than the anterior (fig. 16). The propodos of the latter is oval, of the former oblong quadrangular in form.

The anterior pairs of pereiopoda are normally developed.

The 3 posterior pairs, on the other hand, are comparatively short and have their outer part supplied with dense fascicles of slender bristles. The basal joint of the antepenultimate pair is oval in form; that of the penultimate pair somewhat larger, and narrowed distally. The last pair (fig. 18) have the basal joint still larger, and of a somewhat irregular oval form, with the posterior edge strongly curved in the middle, and densely setiferous.

The anterior pairs of uropoda (fig. 19) are of normal structure, and less unequal than in the preceding species.

The last pair of uropoda (fig. 20) are not nearly so robust as in that species, and have the outer ramus somewhat attenuated distally, exhibiting a distinct, though rather small terminal joint. It is provided on the outer edge with 4, on the inner with 3 fascicles of slender spines accompanied by a few bristles. The inner ramus is rather small, with a single apical bristle.

The telson (fig. 21) is somewhat longer than it is broad, and, as in the preceding species, deeply cleft by a narrow incision. The lateral lobes are obtusely pointed, and carry each 2 lateral, and 2 unequal apical spinules.

Occurrence. — The above-described specimen was taken by Mr. Warpachowsky in the North Caspian Sea, at Stat. 61.

Gen. **Gmelinopsis**, G. O. Sars, n. gen.

Generic Characters. — Body much compressed, more or less distinctly tuberculated laterally, and having the posterior part carinated dorsally. Integuments rather firm. Cephalon with an umboniform prominence on each side. Anterior pairs of coxal plates large and deep. Superior antennæ longer than the inferior, and provided with a small biarticulate accessory appendage. Oral parts nearly as in the genus *Gmelina*. Gnathopoda rather unequal, the anterior ones being the stronger. Last pair of pereiopoda with the basal joint strongly expanded. Uropoda of a structure nearly agreeing with that in the genus *Amathillina*. Telson more or less deeply cleft, with the lateral lobes narrowly pointed.

Remarks. — This new genus is somewhat intermediate between the genera *Gmelina* and *Amathillina*, though apparently more nearly related to the former. It contains as yet 2 species, to be described below.

6. Gmelinopsis tuberculata, G. O. Sars, n. sp.

(Pl. 3, figs 1—19).

Specific Characters. — Cephalon with the lateral prominence obtusely rounded at the tip. Segments of mesosome with distinct lateral tubercles. Last segment of mesosome, and those of metasome elevated dorsally to rounded lamellar expansions. Urosome smooth above. Anterior pairs of coxal plates fully twice as deep as the corresponding segments. Epimeral plates of metasome scarcely produced at the lateral corners. Eyes oblong oval placed close to the anterior edges of the cephalon. Superior antennæ nearly twice as long as the inferior, flagellum longer than the peduncle. Anterior gnathopoda with the propodos rather large, oblong oval, palm oblique and defined below by an obtuse corner carrying a strong spine. Posterior gnathopoda with the propodos much smaller and quadrangular in form. Last pair of pereiopoda with the basal joint obliquely expanded, nearly cordiform in shape. Telson cleft nearly to the base, the cleft gradually widening posteriorly, lateral lobes tapering conically, and armed each with a slender apical spine, accompanied by a small hair, and with a delicate lateral bristle. Length of adult female 8 mm.

Remarks. — This form may be regarded as the type of the genus, and is easily distinguished from the succeeding species by the different form of the lateral prominences of the cephalon and of the dorsal expansions, as also by the structure of the telson.

Description. — The length of a fully adult female specimen measures about 8 mm.

The body is highly compressed and, seen laterally (fig. 1), rather stout, with the back considerably curved, and distinctly carinated in its posterior part. All the segments of the mesosome have their lateral parts distinctly prominent, forming a series of well-defined lateral tubercles. Moreover, the last segment of the mesosome, and those of the metasome are elevated to laminar dorsal expansions, which, however, are not very prominent, and are obtusely rounded.

The cephalon (see fig. 2) is comparatively short, but little longer than the 1st segment of the mesosome, and exhibits in front a distinct, though not very prominent rostral projection. The lateral lobes are rather small and nearly rectangular. Behind them issues, from each side of the head, a very conspicuous umboniform prominence, which extends obliquely downwards, and is obtusely rounded at the tip.

The anterior pairs of coxal plates are rather large, fully twice as deep as the corresponding segments, and have the distal edge slightly crenulated, and fringed with short bristles. The 1st pair (see fig. 11) are but little expanded distally, and are rounded at the tip. The 2 succeeding ones (see fig. 12) are oblong quadrangular in shape, being transversely truncated at the tip. The 4th pair are but little broader than the preceding ones, and exhibit posteriorly a slight emargination, defined below by an obtuse angle.

The 3 posterior pairs of coxal plates are comparatively small, and rapidly diminish in size, the antepenultimate pair (see fig. 14) having the anterior lobe considerably deeper than the posterior.

The epimeral plates of the metasome are well-developed, but have all the lateral corners rounded off.

The urosome is rather short and stout, being perfectly smooth, without any dorsal spines or hairs.

The eyes (see fig. 3) are of moderate size, and oblong oval in form. They are slightly protuberant, and are placed close to the anterior edges of the head.

The superior antennæ (fig. 3) equal about $\frac{1}{4}$ of the length of the body, and are rather slender, with the 1st joint of the peduncle fully as long as the other 2 combined. The flagellum is somewhat longer than the peduncle, and composed of about 12 articulations. The accessory appendage is extremely small, though, on a closer examination (fig. 3 a), it is found to consist of 2 well-defined articulations, the outer of which, however, is so very minute, as easily to escape attention.

The inferior antennæ (fig. 4) are of very inconsiderable size, being scarcely more than half as long as the superior, and have the last joint of the peduncle much smaller than the penultimate one. The flagellum scarcely exceeds in length the last peduncular joint, and is only composed of 4 articulations.

The oral parts (figs. 5—10) on the whole agree in their structure with those in the genus *Gmelina*, and need not therefore be described in detail.

The gnathopoda are rather unequally developed, the anterior ones (fig. 11) being much more powerful than the posterior (fig. 12). The propodos of the former (fig. 11) is very large, and oblong oval in form, with the palm rather oblique, and defined below by an obtuse corner, carrying a strong spine. The propodos of the posterior gnathopoda (fig. 12) is scarcely half as large, and is of oblong quadrangular form, with the palm nearly transverse.

The pereiopoda are rather slender and edged with fascicles of delicate bristles. The 2 anterior pairs (fig. 13) are of quite normal structure. Of the posterior pairs the penultimate one (fig. 15) is the longest, and has the basal joint somewhat larger than in the antepenultimate pair (fig. 14), though exactly similar in shape, both being oblong oval and somewhat narrowed distally. The last pair (fig. 16) are prominently distinguished by the large size of the basal joint, which is greatly expanded, and almost cordiform in shape, its greatest breadth occurring in its distal part. The strongly curved edge of the expansion is minutely crenulated, and fringed with short bristles.

The 2 anterior pairs of uropoda (fig. 17) are normally developed, with the rami subequal, and armed at the tip with a dense bundle of strong spines.

The last pair of uropoda (fig. 18) are comparatively short and thick, extending but little beyond the others, and on the whole resembling those in the genus *Amathillina*. As in that genus, the outer ramus is rather thick, and slightly curved, with a small terminal joint, and carrying on each side a limited number of fasciculated spines, accompanied by a few slender bristles. The inner ramus is extremely small and scale-like.

The telson (fig. 19) is rather narrow, and scarcely at all attenuated distally. It is cleft nearly to the base by a rather wide angular incision, and has the lateral lobes conically tapered, and each provided at the tip with a slender spine accompanied by a minute hair. Moreover, at some distance from the tip exteriorly, a delicate bristle is affixed.

Occurrence. — Of this form, solitary specimens were taken by Mr. Warpachowsky at Stat. 63 and 70, in the eastern part of the North Caspian Sea. Another specimen is contained in the collection of Dr. Grimm, having been taken in the southern part of the Caspian Sea, at Stat. 29, from a depth of 28 fathoms.

7. Gmelinopsis aurita, G. O. Sars, n. sp.

(Pl. 3, figs. 20—28).

Specific Characters. — Cephalon with the lateral prominences greatly produced, spiniform. Lateral tubercles of mesosome less prominent than in the preceding species. The last 2 segments of mesosome, and those of metasome elevated to well-marked lamellar expansions, the last one rounded, the others triangular. Coxal plates about as in *G. tuberculata*, though somewhat less deep. The 2 posterior pairs of epimeral plates of metasome nearly rectangular. Urosome with small hairs dorsally, and a very minute denticle on each side of last segment. Eyes oblong oval. Superior antennæ not nearly twice as long as the inferior, flagellum scarcely attaining the length of the peduncle. Gnathopoda less unequal than in *G. tuberculata*. Pereiopoda and uropoda nearly as in that species. Telson, however, rather different, triangular, considerably narrowed distally, and cleft only in its posterior half. Length 8 mm.

Remarks. — Though very nearly allied to *G. tuberculata*, this form may at once be distinguished by the peculiar development of the lateral prominences of the head, the increased number of the dorsal expansions, and the triangular form of the latter. Moreover, the gnathopoda appear less unequal, and the telson is of a very different shape.

Description. — The length of the solitary specimen examined, which seems to be of female sex, is 8 mm., or about the same as in the preceding species.

The form of the body (see fig. 20) resembles that of *G. tuberculata*, though being, perhaps, still shorter and stouter, with the back very much curved, and very distinctly carinated in its posterior part.

The cephalon (fig. 22) somewhat exceeds in length the 1st segment of the mesosome, and has the lateral lobes very small and obtusangular. On the other hand, the lateral protuberances are greatly developed, forming a pair of spiniform projections extending obliquely forwards, and looking, if the head is viewed from above (fig. 21) like a pair of pointed ears, — hence the specific name.

The lateral tubercles of the mesosome are somewhat less prominent than in the preceding species, though distinct on all the segments of this division. On the other hand, the dorsal expansions are more pronounced, and present on the 2 last segments of the mesosome and those of the metasome, being accordingly 5 in number, whereas in *G. tuberculata* only 4 such expansions are found. Of these expansions, the hindmost is rounded, the other 4 pronouncedly triangular in form.

The urosome is very short and stout, and exhibits dorsally, at the end of each segment, a few small hairs. Moreover, on each side of the last segment, at the base of the telson, occurs (see fig. 23) a small denticle not found in the preceding species.

The anterior pairs of coxal plates exhibit a similar form to that in the preceding species, but are somewhat less deep.

The 2 posterior pairs of epimeral plates of the metasome appear nearly rectangular.

The eyes (see fig. 22) resemble in form and size those in *G. tuberculata*, and, as in that species, are placed close to the anterior edges of the cephalon.

The superior antennæ (ibid.) scarcely equal in length $1/4$ of the body, and have the flagellum somewhat shorter than the peduncle and composed of only 10 articulations. The accessory appendage, as in *G. tuberculata*, is extremely small and biarticulate.

The inferior antennae (ibid.) exceed half the length of the superior, but otherwise agree in structure with those in the preceding species.

The gnathopoda, which however in the solitary specimen could not be examined more closely, appear somewhat less unequal than in *G. tuberculata*, though apparently of a very similar structure.

In the structure of the pereiopoda and uropoda, no essential difference from that found in the preceding species, could be detected.

The telson (fig. 23), on the other hand, looks very different. It is of a triangular form, being about as long as it is broad at the base, and has its outer part conically tapered. The cleft is very narrow and confined to only the outer half of the telson. The lateral lobes are pointed, and each armed at the tip with a minute spine accompanied by 2 small hairs.

Occurrence. — The above-described specimen was taken by Dr. Grimm in the southern part of the Caspian Sea, from the considerable depth of 108 fathoms.

Gen. **Gammaracanthus**, Sp. Bate.

8. **Gammaracanthus caspius**, Grimm. MS.

(Pl. 4, figs. 1—6).

Specific Characters. — Dorsal carina distinct throughout, the anterior segments of the mesosome also being elevated to well-defined lamellar expansions, that of 1st segment rounded, the others triangular and successively increasing in size, those of metasome and urosome sharply pointed. Rostrum but very slightly curved, and extending to the end of the basal joint of the superior antennae. Coxal plates nearly as in *G. relictus*; epimeral plates of metasome acutely produced at the lateral corners. Eyes oval reniform, with a well-marked sinus anteriorly. Antennae nearly as in *G. relictus*. Anterior gnathopoda with the propodos scarcely as large as that of the posterior ones, and much less elongated than in *G. relictus*. Pereiopoda comparatively less slender than in that species, outer part of the 3 posterior pairs densely clothed with short spines. Uropoda and telson of the usual structure. Length of adult female 36 mm.

Remarks. — The present form is nearly allied to *G. relictus* G. O. Sars, but exhibits some well-marked differences, so as more properly to be regarded as a distinct species. Both forms, however, are in all probability originally descended from the arctic species, *G. loricatus* (Sabine); indeed, the present species exhibits in some points a closer relationship to that form than does *G. relictus*.

Description. — The length of the largest specimen examined, an adult, ovigerous female, measures no less than 36 mm., and this form accordingly is of very considerable size, though not nearly as large as the arctic species.

The body (see fig. 1) exhibits the form characteristic of the genus, being somewhat slender, and distinctly carinated both dorsally and laterally.

The dorsal expansions in this form, unlike what is the case in *G. relictus*, are well defined also in the anterior segments of the mesosome, though here somewhat lower than in the posterior segments. That of the 1st segment is evenly rounded, whereas the others are triangular and, especially on the metasome, sharply pointed. The expansions of the 2 anterior segments of the urosome even assume a spiniform character. The lateral keels are distinct throughout the whole mesosome and metasome, and are even continued along the sides of the 1st segment of the urosome.

The cephalon is produced in front to a rather long, sharply-pointed rostrum, which, however, is but very slightly curved, and does not extend beyond the basal joint of the superior antennæ. In the arctic species the rostrum is much stronger, whereas in *G. relictus* it does not extend nearly to the end of the basal joint of the superior antennæ. The lateral lobes are scarcely at all produced, and are quite evenly rounded. Behind them, there is on each side of the head a small umboniform protuberance.

The anterior pairs of coxal plates are a little deeper than the corresponding segments, and oblong quadrangular in form, with an obtuse keel running along their outer face. They successively increase in size posteriorly, the 4th pair being nearly 3 times as large as the 1st, and having the posterior edge slightly concave.

Of the posterior pairs of coxal plates, the last one is quite simple, whereas the 2 anterior pairs are each divided into 2 triangular lappets. In the antepenultimate pair the anterior lappet is very large, being almost as deep as the preceding pair.

The epimeral plates of the metasome are rather large, and all of them produced at the lateral corners to an acute point.

The urosome is of moderate size, and has the last segment quite simple, whereas the 2 anterior ones, as above stated, are elevated to strong, almost spiniform dorsal expansions.

The eyes are oval reniform in shape, having anteriorly a distinct sinus, not found in *G. relictus*. The pigment is very dark.

The superior antennae are rather slender, and only clothed with very small hairs. They somewhat exceed in length $1/3$ of the body, and have the peduncle rather elongated, though considerably shorter than the flagellum, which latter is divided into numerous short articulations. The accessory appendage (see fig. 2) does not attain the length of the last peduncular joint, and is composed of 4 articulations.

The inferior antennæ are very feeble in structure, and scarcely half as long as the superior. The basal joint is globularly tumefied, and the flagellum about as long as the last peduncular joint.

The gnathopoda (figs. 3, 4) are powerfully developed, and exhibit the structure characteristic of the genus, though differing, as to the relative proportion of the propodos, from both the other species. In the anterior pair (fig. 3) the propodos is strongly expanded distally, and almost triangular in form, but is comparatively much less elongated than in *G. relictus*, the greatest breadth considerably exceeding half the length. The palm is evenly arcuate, and somewhat longer than the hind margin, from which it is defined by an obtuse corner carrying several strong spines, one of which is much elongated. The propodos of the posterior gnathopoda (fig. 4), unlike what is the case in *G. relictus*, is somewhat larger than that of the anterior ones, and exhibits a rather different form, being broadest at the base, and gradually tapering distally. The palm is very oblique, occupying almost the whole inferior edge, and is defined behind by an obtusely rounded prominence carrying a rather large number of strong spines arranged in 2 bundles.

The pereiopoda appear on the whole less slender than in *G. relictus*, more resembling those in the arctic species. The 2 anterior pairs are, as usual, much smaller than the 3 posterior, and are edged with fascicles of delicate bristles (see fig. 5). Of the posterior pairs, the 2 anterior ones are much elongated, even exceeding in length half the body, and have their outer part closely edged with short spines. The basal joint of both these pairs is rather narrow, oblong oval, and provided with a slight keel running along the outer surface. The last pair are considerably shorter than the 2 preceding ones, and have the basal joint somewhat larger and more expanded in its proximal part.

The uropoda and telson would seem to be constructed in the very same manner as in the other 2 species.

Occurrence. — Of this pretty form, some specimens were collected by Dr. Grimm in the southern part of the Caspian Sea, from a depth of 108 fathoms.

Gen. **Amathillina**, Grimm.

Of this genus, established by Dr. Grimm, and apparently peculiar to the Caspian Sea, the present author has described, in his 1st article on the Amphipoda, 2 species: *A. cristata* and *affinis*. Three other species are now added, increasing the number of species to 5 in all.

9. **Amathillina spinosa**, Grimm.

(Pl. 4, figs. 7—16).

Amathillina cristata, var. *spinata*, Grimm. MS.

Specific Characters. — Body agreeing in form with *A. cristata*, but having the back distinctly carinated throughout, all the segments of mesosome and metasome being elevated to well-defined triangular expansions increasing in size posteriorly, the hindmost not differing in shape from the others. First segment of urosome with a small rounded expansion dorsally, the other 2 carrying a number of minute hairs at the hind edge. Cephalon with the lateral lobes somewhat prominent, and transversely truncated at the tip. Coxal and epimeral plates about as in *A. cristata*. Eyes reniform and somewhat obliquely disposed. Antennæ and gnathopoda nearly agreeing in structure with those parts in *A. cristata*. Pereiopoda rather slender, and having their outer part densely clothed with fascicles of slender spines, basal joint of the 3 posterior pairs in both sexes less expanded than in the type species. Uropoda and telson not differing essentially from those parts in *A. cristata*. Length of adult male 25 mm.

Remarks. — This form would seem to have been regarded by Dr. Grimm as only a variety of *A. cristata*, since most of the specimens have been labelled *A. cristata*, var. *spinata*[1]. In my opinion, it ought, however, to be regarded as a distinct, though nearly-allied species, differing, as it does, from the typical form, not only in its much larger size, but also in some structural details, for instance, in the much more fully developed dorsal crest, and in the hindmost dorsal expansion being triangular like the preceding ones. Moreover, the basal joint of the 3 posterior pairs of pereiopoda is in both sexes much less expanded than in *A. cristata*.

Description. — Adult male specimens attain a length of 25 mm., and this form accordingly grows to a considerably larger size than the type species.

The general form of the body (see fig. 7) resembles that in *A. cristata*, being moderately slender and somewhat compressed, with the back more or less curved. The dorsal crest is, however, much more fully developed, and extends over a greater part of the body, all the segments of both the mesosome and the metasome being elevated to well-defined triangular dorsal expansions, successively increasing in size posteriorly, and even the 1st segment of the urosome exhibits traces of a dorsal crest, being produced in its posterior part to a small, but well-defined rounded expansion. The expansion of the last segment of the metasome, which in *A. cristata*

---

[1] This adjective form is scarcely acceptable, and should more properly be *spinosa*.

always differs conspicuously from the others by its rounded, gibbous shape, is in the present species of the very same appearance as those preceding it.

The cephalon about equals in length the first 2 segments of the mesosome combined, and exhibits a form similar to that in the type species, the lateral lobes being somewhat prominent and transversely truncated at the tip.

The coxal and epimeral plates are likewise of much the same appearance as in that species.

The urosome, however, as above stated, differs in the fact of its 1st segment having a well defined, though rather small dorsal expansion, of which no trace is found in *A. cristata*. The last 2 segments have dorsally a few small hairs, but are destitute of any true spinules.

The eyes are pronouncedly reniform in shape, and somewhat obliquely disposed, with dark pigment.

The superior antennæ exhibit the slender form characteristic of the genus, and are perhaps a little more elongated than in *A. cristata*. The 2nd joint of the peduncle is nearly as long as the 1st, though considerably narrower, and the 3rd joint is about half its length. The flagellum is almost twice as long as the peduncle, and divided into numerous short articulations. The accessory appendage (see fig. 8) is fully as long as the last peduncular joint, and in the male is composed of 6 articulations.

The inferior antennæ are considerably shorter than the superior, and have the last 2 joints of the peduncle nearly of equal length. The flagellum about equals in length those joints combined.

The gnathopoda, as in the other species, are very different in the 2 sexes, being in the female (figs. 11, 12) rather small and feeble, whereas in the male (figs. 9, 10) they are very powerfully developed, with the propodos of considerable size. In structure, these limbs nearly agree with those in *A. cristata*.

This is also the case with the pereiopoda, though, on a closer comparison, the 3 posterior pairs are found to differ in the fact of the basal joint being comparatively narrower and of nearly similar shape in the 2 sexes (see figs. 7, 13, 14).

The last pair of uropoda (fig. 15) are very short and thick, scarcely at all extending beyond the others, and having the outer ramus but little longer than the basal part, and only armed with 3 short spines and a few delicate hairs; its terminal joint is well defined and spiniform. The inner ramus, as in the other species, is extremely small and scale-like.

The telson (fig. 16) is short and broad, and, as usual, cleft to the base, the cleft somewhat widening behind. Each of the lateral halves is only provided with 2 small apical hairs, and has moreover a slender bristle exteriorly.

Occurrence. — Numerous specimens of this beautiful form were taken by Dr. Grimm in the southern part of the Caspian Sea, from the considerable depth of 108 fathoms.

10. Amathillina Maximowiczi, G. O. Sars, n. sp.

(Pl. 5, figs. 1—14).

Specific Characters. — Body somewhat less slender than in the preceding species, with the mesosome not at all carinated dorsally. Each of the segments of metasome elevated to a comparatively low dorsal expansion, that of last segment not differing from the other 2. Lateral lobes of cephalon very short, and obtusely truncated at the tip. Anterior pairs of coxal plates comparatively small. The last 2 pairs of epimeral plates of metasome slightly produced at the lateral corners. Urosome without any dorsal projection. Eyes oblong oval. Superior antennæ fully twice as long as the inferior, accessory appendage triarticulate. Gnathopoda of the usual structure. Pereiopoda comparatively short and stout, and having their outer part edged with fascicles of slender bristles; basal joint of penultimate pair unusually broad, that of last pair expanded at the infero-posteal corner to a very prominent rounded lobe extending in female to the middle of the meral joint. Last pair of uropoda with the outer ramus almost twice as long as the basal part. Telson less broad than in the preceding species, each half with 2 slender apical spines. Length of adult female scarcely exceeding 6 mm.

Remarks. — This form is at once distinguished from the other known species by the very slight development of the dorsal crest, which is only confined to the metasome. Moreover, the short and stout form of the pereiopoda, and the shape of the basal joint of the posterior pairs may serve for easily recognizing the species. I have much pleasure in dedicating this form to its discoverer Mr. Maximowicz.

Description. — The length of fully adult female specimens scarcely exceeds 6 mm., that of male specimens is, as usual, somewhat more. But this form is in every case far inferior in size to the 3 previously described species.

The form of the body (see fig. 1), especially in the female, is rather short and stout, and somewhat tumid in the anterior part. The dorsal crest, so highly developed in most other species, is in this form confined to the metasome, and is there rather low, each of the 3 segments having a rather slight dorsal expansion fringed with small hairs. The hindmost expansion is somewhat less prominent than the other 2, but of an exactly similar triangular form.

The urosome has none of the segments produced dorsally, and is clothed with a few small hairs.

The cephalon does not quite attain the length of the first 2 segments of the mesosome combined, and has the lateral lobes very short and blunted at the tip.

The anterior pairs of coxal plates are comparatively small, being but little deeper than the corresponding segments, and have their distal edge fringed with scattered bristles. The 1st pair (see fig. 4) are slightly expanded in their outer part, which is evenly rounded off. The 4th pair (fig. 6) have the posterior expansion transversely truncated, and edged with 5 bristles. The posterior pairs are normally developed.

The last 2 pairs of epimeral plates of the metasome are considerably larger than the 1st, and have the lateral corners slightly produced.

The eyes are of moderate size, and are oblong oval in form, without any sinus anteriorly.

The superior antennae (fig. 2) somewhat exceed half the length of the body, and have the 1st joint of the peduncle much thicker, and also longer than the 2nd. The flagellum is very slender, being about twice as long as the peduncle, and is composed of about 20 articulations. The accessory appendage (see fig. 3) does not attain the length of the last peduncular joint, and is 3-articulate.

The inferior antennae are rather poorly developed, being in female scarcely half as long as the superior. In the male they are, as usual, somewhat longer.

The gnathopoda exhibit the usual structure, being rather small in the female (figs. 4, 5), whereas in the male (figs. 10, 11) they are much stronger, with the propodos large and somewhat expanded distally, the palm being in both pairs somewhat oblique and defined below by an obtuse corner, armed with several strong spines.

The pereiopoda are comparatively short and stout, and have their outer part edged with a restricted number of slender, fasciculated bristles. The basal joint of the antepenultimate pair (fig. 7) exhibits the usual oval form; that of the penultimate pair (fig. 8), on the other hand, especially in the female, is unusually broad, being rounded quadrangular in form, with the posterior expansion evenly curved. The basal joint of the last pair (fig. 9) is, as usual, still larger, and obliquely expanded, so as to form a greatly projecting rounded lobe, extending in the female even to the middle of the meral joint. In the male (see fig. 12) however, this lobe is somewhat less prominent.

The last pair of uropoda (fig. 13) extend a little beyond the others, and have the outer ramus almost twice as long as the basal part, being otherwise of the usual structure.

The telson (fig. 14) is somewhat less broad than in the preceding species, and has the cleft narrower. Each of the two halves carries on the tip 2 slender spinules, and has, moreover, exteriorly, at some distance from the tip, a delicate hair.

Occurrence. — Several specimens of this form, males and females, were collected by Mr. Maximowicz in the bay Karabugas.

11. Amathillina pusilla, G. O. Sars, n. sp.

(Pl. 5, figs. 15—25).

Specific Characters. — Body, especially of female, very short and stout, with the last 2 segments of mesosome, and those of metasome elevated to very prominent dorsal expansions, the hindmost being rounded, the other 4 triangularly pointed. Cephalon comparatively larger than in the other species, but of a similar form. Coxal plates comparatively small. Eyes oval reniform. Superior antennae very slender and elongated, being twice as long as the inferior, accessory appendage very small. Gnathopoda in female somewhat unequal, the propodos of the anterior ones being considerably larger than that of the posterior, which is very narrow, oblong quadrangular; those of male powerfully developed and nearly equal-sized. Pereiopoda rather strongly built, and comparatively more elongated than in *A. Maximowiczi*, basal joint of penultimate pair not particularly broad, that of last pair obliquely expanded, forming a narrowly-rounded projecting lobe at the infero-posteal corner. Last pair of uropoda rather short, outer ramus with no lateral spines. Telson with the cleft very narrow, lateral halves obtusely truncated at the tip, and provided with only a single very minute apical spinule. Length of adult female 4 mm.

Remarks. — The present new species is nearly allied to *A. cristata*, of which I formerly believed it to be only a variety. Having, however, submitted the animal in both sexes to a careful anatomical examination, I am now of the opinion that this form ought to be regarded as specifically distinct, since it differs conspicuously from the type species, not only in its very inferior size, but also in some structural details mentioned in the above diagnosis.

Description. — The length of fully adult, ovigerous females scarcely exceeds 4 mm., and this form is accordingly much the smallest of the 5 species as yet known.

The body, especially in the female (fig. 15), is exceedingly short and stout, with the back generally much curved. In male specimens it appears somewhat more slender, though still rather robust. The anterior half of the body is perfectly smooth, without any trace of a dorsal crest. The posterior half, on the other hand, comprising the last 2 segments of the mesosome and those of the metasome, is very distinctly crested, each of the segments being produced dorsally to a very prominent lamellar expansion. The 4 anterior expansions are triangularly pointed, whereas the 5th, as in *A. cristata*, is rounded, gibbous.

The urosome is very short and stout, and without any dorsal prominences, being only clothed with small hairs.

The cephalon appears comparatively larger than in the other species, considerably exceeding in length the first 2 segments of the mesosome combined. In shape however, it agrees with that in the other species, and has the lateral lobes bluntly truncated.

The anterior pairs of coxal plates are comparatively small, and have the distal edge nearly smooth. The 1st pair are scarcely at all expanded distally, and are about same breadth as the 2 succeeding ones. The 4th pair exhibit a form similar to that in the preceding species.

The epimeral plates of the metasome are of moderate size, and the 2 posterior pairs nearly rectangular.

The eyes are oval reniform, having a slight sinus anteriorly.

The superior antennae are very slender and elongated, exceeding in length even $\frac{2}{3}$ of the body, and having the 1st joint of the peduncle rather large, being almost as long as the other 2 combined. The flagellum is fully twice as long as the peduncle, and is composed of about 16 articulations. The accessory appendage is rather small, and composed, in the female (fig. 16) of only 2, in the male (fig. 17) of 3 articulations.

The inferior antennae are about half the length of the superior, and of the usual structure.

The gnathopoda in the female are somewhat more unequal than in the other species, the anterior ones (fig. 18) being considerably more strongly built than the posterior (fig. 19), though less elongated. The propodos of the former (fig. 18) is rather broad, oval in form, and has the palm somewhat oblique; that of the latter (fig. 19) is unusually narrow, oblong linear, with the palm very short and almost transverse. In the male, the gnathopoda (figs. 26, 27) are, as usual, very powerfully developed and nearly equal-sized, the propodos being in both pairs large and tumefied.

The perciopoda (figs. 20—23) are rather strongly built, and comparatively more elongated than in *A. Maximowiczi*. Their outer part is edged with fascicles of bristles intermingled with spines, and in all of them the dactylus is very strong and curved. The basal joint of the penultimate pair (fig. 22) is not particularly broad, but is of a form similar to that of the antepenultimate pair (fig. 21), though somewhat larger. The basal joint of the last pair (fig. 23), on the other hand, is considerably expanded, forming at the infero-posteal corner a projecting lobe, which is narrowly rounded at the tip. The hind edge of the joint is distinctly serrated, with extremely small bristles springing from the serrations. In the male, this joint (see fig. 28) appears somewhat less expanded, though of a similar form.

The last pair of uropoda (fig. 24) are short and stout, with the basal part very thick, and armed at the end with strong spines. The outer ramus is but little longer than the basal part, and does not exhibit any lateral spines, whereas a few such spines, intermingled with slender bristles, issue at the base of the small terminal joint. The inner ramus exhibits the usual scale-like appearance.

The telson (fig. 25) is considerably broader than it is long, and has the cleft rather narrow. The lateral lobes are obtusely truncated at the tip, and each carry but a single very minute apical spinule.

Occurrence. — Several specimens of this form are contained in the collection of Dr. Grimm, having been collected partly in the bay of Baku, partly in that of Balchansky. A few specimens were also collected last year by Mr. Andrussow in the bay of Krasnovodsk. It would seem everywhere to be a sub-littoral species.

Gen. **Gammarus**, Lin.

This genus would seem to be very abundantly represented in the Caspian Sea, and exhibits a very remarkable extent of variability as to the specific characters. To the 11 species previously described by the present author, are now added 10 new ones, increasing the number of species to no less than 21 in all, and it is very probable that we are still far from having become acquainted with all the existing forms.

12. **Gammarus Grimmi**, G. O. Sars.

(Pl. 6, figs. 1—10).

Gammarus robustus Grimm, not Smith.

Specific Characters. — Body rather slender and compressed, with the mesosome and metasome perfectly smooth. Urosome having the 2 anterior segments each produced dorsally to a very prominent tubercle tipped with spines; 1st segment with 2 unequal lateral spines; last one with 3 such spines. Lateral lobes of cephalon somewhat prominent, and obtusely truncated at the tip. Coxal plates not very large, somewhat deeper in female than in male. Last pair of epimeral plates of metasome acutely produced at the lateral corners. Eyes oblong oval. Superior antennæ much longer than the inferior, and very slender, flagellum fully twice the length of the peduncle, accessory appendage rather elongated, and composed of 5—7 articulations. Gnathopoda rather unequally developed, the posterior ones being much the stronger, and in male exceedingly powerful, with the propodos large and tumefied. Pereiopoda comparatively more slender than in most other species, basal joint of the 3 posterior pairs rather narrow, that of last pair oblong quadrangular in form, being produced at the infero-posteal corner to a short narrowly rounded lobe. Last pair of uropoda very much elongated, outer ramus narrow, sublinear and densely setous, inner ramus small, scale-like. Telson with the lateral lobes conically tapering, and each armed with 2 apical spines. Length of male 27 mm.

Remarks. — The present form was labelled by Dr. Grimm *G. robustus*; but, as this name has been already appropriated by Prof. S. Smith for an American species, I have seen fit to change it, and to name the species after its discoverer. It is very nearly allied to the form described by the present author in his first article on the Caspian Amphipoda as *G. hæmobaphes* Eichwald, and at first I was indeed inclined to regard it only as a variety of that species. On a closer examination, however, I have found it to differ rather conspicuously in some particulars, so as more properly to be regarded as specifically distinct.

Description. — The length of the largest specimens reaches to 27 mm., and this form accordingly grows to a considerably larger size than *G. hæmobaphes*. As usual, male specimens are generally larger than female ones.

The body (see fig. 1), as compared with that of *G. hæmobaphes*, appears rather more slender and compressed, and has the mesosome and metasome quite smooth throughout. The urosome, on the other hand, has each of its first 2 segments produced dorsally to a very prominent narrow tubercle of a similar shape to that found in *G. caspius*. The tubercle of the 1st segment (see fig. 3) is somewhat stronger than that of the 2nd, and carries on the transversely truncated tip, 4 spines arranged in pairs, and accompanied by a few small hairs. The posterior tubercle has only 2 juxtaposed apical spines. Moreover, the 1st segment has 2 unequal lateral spinules, and 3 similar spines occur on each side of the last segment, at the base of the telson.

The cephalon scarcely attains the length of the first 2 segments of the mesosome combined, and has the lateral lobes rather broad and somewhat prominent, with the tip obtusely truncated.

The anterior pairs of coxal plates resemble in shape those in *G. hæmobaphes*, being somewhat deeper in the female than in the male.

The epimeral plates of the metasome are well developed, and the last pair acutely produced at the lateral corners.

The eyes are of moderate size, oblong oval in form, and somewhat obliquely disposed. The pigment is generally dark, but in some specimens it appeared considerably lighter.

The superior antennæ are very slender and elongated, exceeding half the length of the body, and with the joints of the peduncle successively diminishing in size. The flagellum is fully twice as long as the peduncle, and is divided into numerous short articulations. The accessory appendage (see fig. 2) is very narrow, thread-like, and fully twice as long as the last peduncular joint. It is composed of 7 articulations of equal length.

The inferior antennæ are considerably shorter than the superior, but more strongly built, especially in the male, and are provided with dense fascicles of bristles along the posterior edge of both the peduncle and the flagellum.

The gnathopoda exhibit a similar structure to that in *G. hæmobaphes*, and, as in that species, are rather unequally developed, the posterior being much the stronger. In the female this pair (fig. 7) have the propodos nearly twice as large as in the 1st pair (fig. 6), though exhibiting a similar form, the palm being somewhat oblique, and defined below by an obtuse corner armed with a number of strong spines. In the male both pairs are rather unlike, and still more unequal in size than in the female. The anterior ones (fig. 4) have the propodos of moderate size, and oblong oval in form, with the palm very oblique. In the posterior ones (fig. 5) the propodos is of very considerable size, and strongly tumefied in its proximal part, having numerous dense fascicles of bristles along the inferior edge. The palm is obliquely arcuate, and defined below by a rather projecting corner armed with 3 strong spines. The dactylus in both pairs is much coarser than in the female and strongly curved.

The pereiopoda are comparatively much more slender than in *G. hæmobaphes*, and have their outer part edged with fascicles of short spines. The basal joint of the 3 posterior pairs is conspicuously narrower than in the said species, though exhibiting a similar mutual relationship as to size. In the last pair (fig. 8) this joint is of an oblong quadrangular form, with the infero-posteal corner produced to a short, narrowly rounded lappet; its posterior edge is nearly straight and very distinctly serrated.

The last pair of uropoda (fig. 9) are very much elongated, being fully as long as the urosome, and having the outer ramus narrow, sub-linear, and densely fringed all round with setæ. The inner ramus is very small and scale-like.

The telson (fig. 10) has the lateral lobes obtusely pointed, and each armed on the tip (see fig. 10*a*) with 2 short spines accompanied by some delicate hairs.

Occurrence. — This form has been collected in great numbers by Dr. Grimm, in the middle and southern parts of the Caspian Sea, the depth varying from 35 to 108 fathoms.

13. Gammarus subnudus, G. O. Sars. n. sp.

(Pl. 6, figs. 14—19).

Specific Characters. — Body somewhat tumid and quite smooth throughout. Cephalon with the lateral lobes rather broad and somewhat obliquely truncated at the tip, the inferior corner being the more prominent. Anterior pairs of coxal plates comparatively large and broad, especially the 4th pair. The last 2 epimeral plates of metasome nearly rectangular. Urosome without any spines or projections dorsally. Eyes oblong oval. Superior antennæ comparatively short, though a little longer than the inferior, flagellum about the length of the peduncle, accessory appendage rather small, 3-articulate. Gnathopoda in female rather unequal, the posterior ones being much the stronger. Pereiopoda not very elongated, and almost naked, basal joint of antepenultimate pair subquadrangular, that of last pair considerably expanded, and forming at the infero-posteal corner a rounded projecting lobe, posterior edge but slightly crenulated. Last pair of uropoda of moderate size, outer ramus rather broad, flattened, setous all round; inner ramus small, scale-like. Telson with the lateral lobes comparatively broad and blunted at the tip, each with a single small apical spinule. Length of adult female 8 mm.

Remarks. — This new species may easily be recognized by its comparatively tumid body, the smooth urosome, and the unusually slight spinous armature of the pereiopoda, which latter character has given rise to the specific name.

Description. — The length of fully adult, ovigerous specimens does not exceed 8 mm.

The body (see fig. 14) appears rather robust and somewhat tumid, with the back broadly rounded, and quite smooth throughout.

The cephalon is but little longer than the 1st segment of the mesosome, and has the lateral lobes rather broad and somewhat obliquely truncated, with the inferior corner the more prominent.

The anterior pairs of coxal plates are comparatively large and broad, being considerably deeper than the corresponding segments. The 1st pair are not at all expanded distally and, like the 2 succeeding ones, obtusely truncated at the tip, without any marginal bristles. The 4th pair are of considerable breadth, and have the posterior expansion transversely truncated.

The epimeral plates of the metasome are of moderate size, and the last 2 pairs nearly rectangular.

The urosome is quite smooth, without any spines or dorsal projections, having only at the end of each segment a few small hairs.

The eyes are of middle size and oblong oval in form, with dark pigment.

The superior antennæ are not very much elongated, scarcely exceeding in length $1/3$ of the body, and with the joints of the peduncle successively diminishing in size. The flagellum does not exceed the peduncle in length, and is composed of about 16 articulations. The accessory appendage is comparatively small, not attaining the length of the last peduncular joint, and is only composed of 3 articulations.

The inferior antennæ are a little shorter than the superior, and are somewhat more densely setiferous. The flagellum does not attain the length of the last 2 peduncular joints combined.

The gnathopoda of the female, as in the preceding species, are rather unequal in size, the posterior ones (fig. 14) being much the stronger, with the propodos more than twice as large as that of the anterior ones (fig. 13). The palm in both pairs is rather oblique, and its defining angle very slight, though armed in the usual manner.

The pereiopoda are not very slender, and are remarkable for their poor supply of spines or bristles, for which reason they at first sight appear nearly naked. The basal joint of the antepenultimate pair (fig. 15) is rounded quadrangular in form, being about as broad as it is long, and having the infero-posteal corner produced to a short rounded lobe. That of the penultimate pair (fig. 16) has the infero-posteal corner not at all produced, and on this account exhibits a more triangular form. The basal joint of the last pair (fig. 17), as usual, is considerably larger than that of the 2 preceding pairs, forming posteriorly a rather broad expansion, which is produced below to a projecting rounded lobe. The posterior edge of this joint is but slightly crenulated, and is provided in its inferior half with a few very small hairs.

The last pair of uropoda (fig. 18) are of moderate size, and have the outer ramus rather broad, sublamellar, and about twice as long as the basal part. It is fringed all round with slender setæ, and has moreover on the outer edge, 2 fascicles of spines. Its terminal joint is rather small, but distinct. The inner ramus, as in most other Caspian species, is very small and scale-like.

The telson (fig. 19) has the lateral lobes rather broad and blunted at the tip, each carrying only a single apical spine accompanied by a small hair.

Occurrence. — Some few specimens of this form are contained in the collection of Dr. Grimm, having been collected in the bay of Baku, from 2—6 fathoms.

14. Gammarus macrocephalus, Grimm. MS.

(Pl. 7, figs. 1—11).

Specific Characters. — Body somewhat robust, with evenly rounded back. Cephalon of quite an unusual size and strongly convex above, lateral lobes short and broad, transversely truncated at the tip. Coxal plates not particularly large. The last 2 pairs of epimeral plates of metasome acutely produced at the lateral corners. Urosome with the 2 anterior segments produced dorsally to very prominent narrow tubercles tipped by spines; no lateral spines on the 1st segment. Eyes very small, reniform. Superior antennæ rather slender, and somewhat longer than the superior, accessory appendage narrow, 6-articulate. Gnathopoda (in male) very unequally developed, the posterior ones being much stronger than the anterior, with the propodos exceedingly large, oval pyriform. Pereiopoda rather slender, and edged with fascicles of short spines, basal joint of last pair moderately expanded and produced at the infero-posteal corner to an obtusely-pointed lappet. Last pair of uropoda rather elongated, though not attaining the length of the urosome, outer ramus narrow, sub-linear, and densely fringed with setæ, between which are scattered spines; inner ramus very small. Telson with the lateral lobes narrowly truncated at the tip, and each carrying 3 apical spines. Length of adult male 24 mm.

Remarks. — This species, established by Dr. Grimm, is at once recognized by the unusual size and peculiar shape of the cephalon, which character, indeed, would seem to have given rise to the specific name proposed by that author. Otherwise it is nearly allied to *G. Grimmi*, exhibiting a rather similar armature of the urosome.

Description. — The length of an adult male specimen measures about 24 mm., and this form accordingly grows to a rather large size.

As compared with *G. Grimmi*, the body appears (see fig. 1) somewhat more robust, with the back broadly rounded and quite smooth.

The cephalon is highly remarkable for its large size and unusual shape. It considerably exceeds in length the first 2 segments of the mesosome combined, and has the dorsal face strongly vaulted, for which reason this species acquires a rather peculiar physiognomy. In its interior numerous strong muscular bundles are seen converging to the buccal area, and the points of insertion for these bundles beneath the integument produce a peculiar areolated appearance of the dorsal face of the head. The lateral lobes are very short and broad, being transversely truncated at the tip. Behind them the inferior edges of the head form a deep emargination encircling the globular basal joint of the inferior antennae.

The anterior pairs of coxal plates are not particularly large, though somewhat deeper than the corresponding segments. The 1st pair are scarcely expanded distally, and are transversely truncated at the tip, whereas the 2 succeeding ones appear more rounded distally. The 4th pair, as usual, are much broader than the others, and form beneath the posterior emargination an angular corner.

The epimeral plates of the metasome are normally developed, the 1st pair being rather small and rounded, whereas the last 2 pairs are comparatively large, and acutely produced at the lateral corners.

The urosome has the 2 anterior segments elevated dorsally, in a manner similar to that in *G. Grimmi*, to narrow prominent tubercles, tipped, the anterior with 4, the posterior with 2 spines. On the sides of the 1st segment no spines are present, but on the last segment, 2 small juxtaposed denticles are found on each side, at the base of the telson.

The eyes are very small, but distinctly reniform, having a rather deep sinus anteriorly.

The superior antennae are rather slender, though somewhat less elongated than in *G. Grimmi*, not attaining half the length of the body. The 1st joint of the peduncle is rather large, equalling in length the other 2 combined. The flagellum does not quite attain to twice the length of the peduncle, and is divided into numerous short articulations. The accessory appendage is nearly twice as long as the last peduncular joint, and extremely narrow, being composed of 6 articulations.

The inferior antennae are somewhat shorter than the superior, but considerably more strongly built, and also more richly supplied with bristles. Their structure is the usual one.

The gnathopoda, at least in the male, are very unequally developed, the posterior ones (fig. 4) being much more powerful than the anterior (fig. 3). The propodos of the former is fully twice as large as that of the latter, and somewhat pyriform in shape, being considerably tumefied in its proximal part, and tapering distally, with the palm rather oblique and defined behind by a slight, spinous corner.

The pereiopoda are rather slender, and have their outer part edged with scattered fascicles of spines (see fig. 5). The basal joint of the antepenultimate pair (fig. 6) does not differ much in shape from that of the penultimate pair (fig. 7), though being somewhat shorter. The basal joint of the last pair (fig. 8) is much larger, forming posteriorly a lamellar expansion, which is produced below to an obtusely-pointed lappet. The hind edge of the joint is but very slightly curved, and is distinctly serrated.

The 2 anterior pairs of uropoda (fig. 10) exhibit the usual structure, the rami being nearly equal-sized, and tipped by numerous short spines.

The last pair of uropoda (fig. 9) are rather elongated, though not quite attaining the length of the urosome. The outer ramus is very narrow, sublinear, and is densely fringed all round with slender setæ, besides having on each side 3 fascicles of short spines. Its terminal joint is so very minute as easily to escape attention. The inner ramus exhibits the usual rudimentary condition.

The telson (fig. 11) is comparatively small, and has the lateral lobes narrowly truncated at the tip, each carrying 3 apical spines accompanied by a few small hairs.

Occurrence. — Some few specimens of this form were collected by Dr. Grimm in the middle part of the Caspian Sea, from a depth of 35 fathoms.

15. Gammarus tenellus, G. O. Sars, n. sp.

(Pl. 7, figs. 12—22).

Specific Characters. — Body extremely slender and compressed, with the mesosome and metasome quite smooth. Cephalon with the lateral lobes very obliquely truncated, so as to form in front an acute corner. Coxal plates not very large. Last pair of epimeral plates of metasome acutely produced at the lateral corners. Urosome distinctly spinulose above, the spinules being upturned, and arranged in dorsal and lateral fascicles. Eyes well-developed, oblong oval. Antennæ rather slender, both pairs being densely fringed with delicate setæ, the superior ones much the longer, with the flagellum twice as long as the peduncle, and the accessory appendage 4-articulate. Gnathopoda in both sexes subequal in size, being in female rather feeble, in male somewhat stronger. Pereiopoda moderately slender, and edged with dense fascicles of bristles, basal joint of last pair scarcely differing either in size or shape from that of the penultimate pair. Last pair of uropoda very much elongated, considerably exceeding in length the urosome, outer ramus sub-linear and edged with dense fascicles of spines, inner ramus very small. Telson small, and abruptly narrowed in its outer part, with the lateral lobes obtusely pointed, and each having 3 apical and 2 lateral spinules. Length of adult female 6 mm.

Remarks. — This new species is easily distinguishable by its slender and compressed body, the conspicuously spinulose urosome, the uniform shape of the basal joint of the 3 posterior pairs of pereiopoda, and finally, by the great development of the last pair of uropoda.

Description. — Fully adult female specimens do not exceed in length 6 mm., male specimens are, as usual, a little larger.

The body (see fig. 12) is exceedingly slender and much compressed, with the mesosome and metasome quite smooth.

The urosome, however, exhibits dorsally a very conspicuous spinulose armature, each segment having 3 fascicles of slender, very much upturned spinules, one median and 2 lateral. Each fascicle contains, as a rule, 2 juxtaposed spinules, sometimes accompanied by a small hair.

The cephalon does not attain the length of the first 2 segments of the mesosome combined, and has the lateral lobes very obliquely truncated, so as to form anteriorly a projecting acute angle.

The anterior pairs of coxal plates are not very large, though a little deeper than the corresponding segments, and they are quite smooth. The 1st pair are scarcely expanded distally, and are obtusely rounded at the tip. The 4th pair (see fig. 16) are not particularly broad, and they have the posterior expansion obtusely truncated.

The epimeral plates of the metasome are well developed, the 1st pair being, as usual, rounded, whereas the 2 other pairs, and especially the last, are acutely produced at the lateral corners.

The eyes are of moderate size, and oblong oval in form, being placed close to the anterior edges of the head.

The superior antennae are very slender, exceeding half the length of the body, and have the peduncle, and partly also the flagellum, clothed posteriorly with slender delicate setae. The 1st joint of the peduncle is, as usual, the largest, though not much longer than the 2nd, whereas the 3rd is very much smaller. The flagellum is fully twice as long as the peduncle, and composed of about 20 articulations. The accessory appendage (see fig. 13) is rather slender, and 4-articulate.

The inferior antennae are considerably shorter than the superior, and, especially in the male, very densely clothed posteriorly with rather long setae. The flagellum is about the length of the peduncle.

The gnathopoda, unlike what is the case in the 3 preceding species, are in both sexes nearly equal-sized. In the female (figs. 14, 15) they are rather feeble in structure, and differ somewhat in the shape of the propodos, which in the posterior ones (fig. 15) appears somewhat more elongated than in the anterior (fig. 14), and has the palm more transverse. In the male both pairs (figs. 23, 24) are considerably larger, though not particularly strong. The propodos of the anterior ones (fig. 23) is oblong in form, with the palm very oblique and somewhat concave, that of the posterior ones (fig. 24) is a little larger, and widens slightly distally, with the palm nearly transverse.

The pereiopoda are moderately slender, and have their outer part edged with fascicles of slender bristles. The 2 anterior pairs (fig. 16) are considerably smaller than the 3 posterior ones (figs. 17—19), which exhibit a very uniform appearance, the basal joint in all of them being oval in shape. That of the last pair (fig. 19), unlike what is generally the case, is only very slightly expanded, and scarcely differs either in size or shape from that of the penultimate pair (fig. 18).

The 2 anterior pairs of uropoda (fig. 21) have the outer ramus a little shorter than the inner, both being tipped by a number of rather slender spines.

The last pair of uropoda (fig. 20) are highly remarkable for their great length, being almost twice as long as the urosome. The outer ramus is rather narrow, and sub-linear, and exhibits on each side 5 fascicles of slender spines. Its terminal joint is well defined, spiniform, and carries a few slender bristles on the tip. The inner ramus is very small, and scale-like.

The telson (fig. 22) is comparatively small and abruptly narrowed in its outer part. The cleft is rather wide, and the lateral lobes obtusely pointed. They are each armed with 3 apical, and 2 lateral spinules, one of which, however, has more properly the character of a bristle.

Occurrence. — Several specimens of this form were collected by Dr. Grimm at Stat. 62, lying at some distance south of Baku, the depth being 6 fathoms.

16. **Gammarus placidus**, Grimm, MS.

(Pl. 8, figs. 1—12).

Specific Characters. — Body extremely slender, though less compressed than in *G. tenellus*. Cephalon with the lateral lobes produced in front to a very acute somewhat deflexed point. Coxal plates not very large. Last pair of epimeral plates of metasome but slightly produced at the lateral corners. Urosome densely spinulose above. Eyes of a rather unusual form, being very narrow, and occupying almost the whole height of the front part of the cephalon, their lower part slightly dilated. Antennae very slender, and densely setiferous, the superior ones longer than the inferior, and having the accessory appendage rather elongated and 7-articulate. Gnathopoda in female rather feeble, and almost equal, propodos of the posterior ones somewhat more elongated than that of the anterior, and having the palm more transverse. Pereiopoda considerably elongated, basal joint of the 3 posterior pairs very narrow, and of uniform appearance in all of them. Last pair of uropoda remarkably elongated, being more than twice as long as the urosome, outer ramus linear and densely clothed with spines, inner ramus rudimentary. Telson small, with the lateral lobes obtusely truncated at the tip, and each armed with a single apical, and a lateral spinule. Length of adult female 13 mm.

Remarks. — This is a very distinct species, and easily recognizable by the slender form of the body, the peculiar form of the eyes, the acutely produced lateral lobes of the cephalon, the slender and densely setous antennae, the narrow form of the basal joint of the 3 posterior pairs of pereiopoda, and finally, the extraordinary length of the last pair of uropoda. In some points it would seem to approach the above-described *G. tenellus.*

Description. — The length of an adult, ovigerous female, not including the last pair of uropoda, measures about 13 mm., and this form accordingly is more than twice as large as its nearest ally, *G. tenellus.*

The body (see fig. 1) is very slender, though less compressed than in the last-named species, and has the mesosome and metasome perfectly smooth.

The cephalon (fig. 2) does not quite attain the length of the first 2 segments of the mesosome combined, and has the lateral lobes produced in front to a rather conspicuous acute point, which is sometimes slightly deflexed.

The anterior pairs of coxal plates are not very large, though somewhat deeper than the corresponding segments. The 1st pair (see fig. 4) are quadrangular in form, the 2 succeeding ones (see fig. 5) a little narrowed in their outer part. The 4th pair are but little broader than the preceding one, and have the posterior expansion rather short.

The epimeral plates of the metasome are of moderate size, the 1st pair being, as usual, rounded, the other 2 but very slightly produced at the lateral corners.

The urosome is conspicuously spinulose above, the spinules being, as in *G. tenellus,* arranged in 3 fascicles on each segment. The lateral fascicles each contain on the 1st segment 4, on the 2nd 3, and on the last 2 spinules.

The eyes (see fig. 2) are highly remarkable by their unusual shape, being extremely narrow, almost band-shaped, and occupying nearly the whole height of the frontal part of the head. Their lower part is slightly dilated and curved anteriorly towards the point of the lateral lobes. The visual elements are well developed, and the pigment very dark.

The superior antennae are very slender and elongated, almost attaining $3/4$ of the length of the body. They are clothed posteriorly with numerous fascicles of slender bristles, and have the 2nd joint of the peduncle fully as long as the 1st. The flagellum is about twice as long as the peduncle, and is composed of numerous short articulations. The accessory appendage is rather fully developed, considerably exceeding in length the last peduncular joint, and is composed of 7 articulations.

The inferior antennae are shorter than the superior, and still more densely clothed with setae, arranged in regular fascicles along the posterior edge of both the peduncle and the flagellum. The latter nearly equals the peduncle in length, and is composed of about 16 articulations.

The gnathopoda of the female (figs. 4, 5) are not very powerful, and are nearly equal-sized, the posterior ones being only a little more elongated than the anterior. Both pairs are densely clothed with slender, fasciculated bristles, and have the propodos slightly different in shape, that of the posterior ones (fig. 5) being somewhat longer and more transversely truncated at the tip. In the male the posterior gnathopoda are considerably larger than the anterior, but as none of the male specimens examined were fully grown, no figures are given here.

The 2 anterior pairs of perciopoda (see fig. 1) are rather slender and somewhat unequal in length, the 1st pair being the longer, but otherwise exactly agreeing with the 2nd pair.

The 3 posterior pairs of perciopoda are more strongly built, and rather elongated, being generally strongly reflexed. The basal joint in all of them (see figs. 7—9) is rather narrow, and in the last pair (fig. 9) not at all differing from that of the penultimate pair, either in size or shape. The outer part of all these legs is edged with scattered fascicles of short spines.

The 2 anterior pairs of uropoda (fig. 10) have both the basal part and the rami rather densely spinous, but are otherwise quite normal in structure.

The last pair of uropoda (fig. 11), on the other hand, are very remarkable from their extraordinary length, being even more than twice as long as the urosome. The outer ramus is of uniform breadth throughout, linear in form, and provided on both edges with numerous fascicles of short spines accompanied by a few small bristles. The terminal joint is spiniform and almost hidden between the spines issuing from the end of the proximal joint. The inner ramus is so extremely small as easily to be overlooked.

The telson (fig. 12) is rather short, being almost twice as broad as it is long, and has the cleft rather narrow. The lateral lobes have the outer edge angularly bent in the middle, and are obtusely truncated at the tip. They carry each a single apical spinule, and a similar lateral one.

Occurrence. — Some few specimens of this form, for the greater part still immature, are contained in the collection of Dr. Grimm, having been collected partly in the bay of Baku, from a depth of 2—6 fathoms, partly in the middle part of the Caspian Sea, from a depth of 40 fathoms. A solitary, quite young, male specimen was taken last year by Mr. Maximowicz in the bay Karabugas.

17. Gammarus platycheir, G. O. Sars, n. sp.

(Pl. 8, figs. 14—27).

Specific Characters. — Body rather robust and tumid, with broadly vaulted back. Lateral lobes of cephalon transversely truncated at the tip. Anterior pairs of coxal plates rather deep, and densely fringed with bristles; 1st pair tapering conically, 2nd pair likewise narrowed distally, 3rd pair oblong quadrangular, 4th pair rather broad, forming beneath the posterior expansion an angular corner. Last pair of epimeral plates of metasome but slightly produced at the lateral corners. Urosome with a few scattered spinules above. Eyes oblong reniform. Antennæ very short, the superior ones not exceeding in length the inferior, and having the flagellum shorter than the peduncle, accessory appendage 4-articulate. Gnathopoda very unequal, the posterior ones being much stronger than the anterior, with the propodos greatly expanded distally, palm arcuate, and having a dense series of submarginal bristles, defining angle very slight, spinous, one of the spines being much elongated, dactylus long, falciform. Pereiopoda rather slender, basal joint of last pair much larger than that of the 2 preceding pairs, oblong oval, hind edge densely setiferous. Last pair of uropoda not much elongated, outer ramus subfoliaceous, fringed with setæ, inner rather small. Telson cleft nearly to the base, lateral lobes obtusely pointed, and each armed with 3 small apical spinules. Length 16 mm.

Remarks. — This is a rather anomalous form, which is only provisionally placed under the genus *Gammarus*, exhibiting, as it does, some points of affinity both to the genus *Niphargoides* and to *Pandorites*.

Description. — The solitary specimen examined, which seems to be of male sex, measures about 16 mm. in length, and is accordingly of rather large size.

The body (see fig. 14) somewhat resembles in form that of the species of the genus *Niphargoides*, being rather robust and tumid, with the back broadly vaulted and quite smooth.

The cephalon is comparatively small, not nearly attaining the length of the first 2 segments of the mesosome combined, and has the lateral lobes rather broad and transversely truncated at the tip.

The anterior pairs of coxal plates are nearly twice as deep as the corresponding segments, and densely fringed with setæ on their distal edge. They are rather unequal both as to size and shape. The 1st pair are much the smallest, and exhibit a somewhat unusual form, being gradually narrowed distally, with the tip obtusely pointed, and extending obliquely forwards. The 2nd pair are somewhat larger than the 1st, and more deflexed, but are likewise considerably narrowed distally. The 3rd pair are much larger, and of the usual oblong quadrangular form. The 4th pair are still larger, nearly as broad as they are deep, and distinctly emarginated posteriorly in their upper part, the emargination being defined below by a projecting corner.

The epimeral plates of the metasome are of moderate size, the 2nd pair being the deepest, and exhibiting in their anterior part a dense fringe of delicate, curved bristles. The last pair are but little produced at the lateral corner, and, like the preceding pair, nearly rectangular.

The urosome has on each of its segments dorsally, 2 small, juxtaposed spinules. On the last segment there is also a single lateral spinule (see fig. 17).

The eyes are of moderate size, and oval reniform in shape, being placed somewhat obliquely.

The superior antennæ are rather short, scarcely exceeding in length the cephalon and the first 2 segments of the mesosome combined, and have the 1st joint of the peduncle about as long as the other 2 taken together. The flagellum (see fig. 15) does not attain the length of the peduncle, and is composed of 12 articulations. The accessory appendage is about $^1/_2$ as long as the flagellum, and 4-articulate.

The inferior antennæ are fully as long as the superior, perhaps even a little longer, and exhibit the usual structure.

The gnathopoda are very unequally developed, the posterior (fig. 16) being much larger than the anterior, and exhibiting a rather anomalous structure somewhat recalling that in the genus *Pandorites*. As in that genus, the carpus is very small, whereas the propodos is of exceedingly large size, and somewhat flattened, being gradually expanded distally, whereby it acquires an almost triangular form. The palm is rather oblique and evenly curved, being provided, somewhat within the edge, with a dense and regular series of strong, spiniform bristles. The defining angle is very slight, and armed with 3 strong spines, one of which is rather elongated. The hind margin of the propodos, which is somewhat shorter than the palm, carries a few fascicles of short bristles, whereas the opposite edge is quite smooth. The dactylus is very long and falciform, its tip being received between the spines of the lower corner, when the dactylus is bent in against the propodos.

The pereiopoda are rather slender, and exhibit on the whole a normal structure. The 2nd pair are somewhat more densely setiferous than the 1st, but otherwise of the very same appearance. The basal joint of the last pair is much larger than that of the 2 preceding pairs, and of oblong oval form, with the posterior edge slightly curved and densely fringed with delicate setæ.

The last pair of uropoda (see fig. 17) are not particularly elongated, though reaching considerably beyond the others, and have the outer ramus rather broad, sub-foliaceous, with a very small terminal joint. It is densely fringed with slender setæ, and has, moreover, on the outer edge, 2 distant spines. The inner ramus is somewhat less rudimentary than in most other species, being nearly $\frac{1}{3}$ as long as the outer, and carries at the tip 2 small spines.

The telson (ibid.) is about as long as it is broad, and scarcely narrowed distally. The cleft does not extend quite to the base, and it gradually widens posteriorly. The lateral lobes are bluntly pointed, and carry each 3 rather small and closely-set apical spines.

Occurrence. — The above-described specimen was taken by Mr. Warpachowsky at Stat. 59, lying at some distance outside the mouth of the Wolga.

18. Gammarus Weidemanni, G. O. Sars, n. sp.

(Pl. 9, figs. 1—11).

Specific Characters. — Body comparatively robust and quite smooth, with broadly vaulted back. Lateral lobes of cephalon but slightly prominent, and obliquely rounded. Anterior pairs of coxal plates rather large, and densely setiferous; 1st pair distinctly expanded in their outer part, 4th pair very broad, with the posterior expansion transversely truncated. Last pair of epimeral plates of metasome but slightly produced at the lateral corners. Urosome quite smooth above. Eyes of moderate size, oval reniform. Antennæ short, equal-sized, the superior ones with the flagellum longer than the peduncle, accessory appendage 5-articulate. Inferior antennæ with the 2 outer joints of the peduncle simple, cylindric, clothed with scattered fascicles of bristles. Mandibular palps of normal size. Gnathopoda in female comparatively strong, and somewhat unequal, the posterior ones being the larger; those in male, as usual, still more powerful. The 2 anterior pairs of pereiopoda very robust, with the meral and carpal joints lamellarly expanded and densely clothed with delicate curved bristles; the 3 posterior pairs likewise rather robust, and having their outer part edged with strong spines accompanied by slender bristles: basal joint of last pair rather large, oval. Last pair of uropoda not much elongated, outer ramus sub-foliaceous and densely setiferous, inner ramus about half the length of the outer. Telson with the lateral lobes rather narrow, and each armed with 3 apical spines. Length of adult male 11 mm.

Remarks. — In its general appearance, this form somewhat resembles *G. aralensis* Uljanin[1]), from which however, it, is easily distinguishable by the perfectly smooth urosome, and by the somewhat different structure of the posterior pairs of pereiopoda. It also comes rather near to *G. abbreviatus* G. O. Sars; but its nearest ally would seem to be *G. mæoticus* Sowinsky, to be described later.

Description. — The length of the largest male specimens measures about 11 mm., and this form is accordingly rather inferior in size to the Caspian variety of *G. aralensis*.

The form of the body (see fig. 1) appears on the whole to be rather robust and somewhat tumid, with the back broadly rounded and perfectly smooth throughout.

The cephalon does not attain the length of the first 2 segments of the mesosome combined, and has the lateral lobes but slightly prominent, and obliquely rounded.

The anterior pairs of coxal plates are rather large, and densely fringed distally with slender bristles. The 1st pair (see fig. 4) are distinctly expanded in their outer part, and accordingly somewhat broader than the succeeding pair (see fig. 5), which, on the contrary, are slightly narrowed distally. The 4th pair are exceedingly broad, with the posterior expansion rather prominent, and transversely truncated at the tip.

The last 2 pairs of epimeral plates of the metasome are of same shape, and but slightly produced at the lateral corners.

The urosome is perfectly smooth above, with only 2 very small juxtaposed spinules on each side of the last segment, at the base of the telson.

The eyes are of moderate size, and are oval reniform in shape, with dark pigment.

The superior antennæ (fig. 2) are rather short, scarcely exceeding in length the cephalon and the first 2 segments of the mesosome combined, and have the 1st joint of the peduncle rather massive, exceeding in length

---

1) This species has been described by the present author in his 2nd article on the Caspian Amphipoda as *G. robustoides*, Grimm.

the other 2 combined. The flagellum somewhat exceeds the peduncle in length, and is composed of about 16 articulations. The accessory appendage is of moderate size, and 5-articulate.

The inferior antennæ (fig. 3) are about same length as the superior, but somewhat more strongly built. The 2 outer joints of the peduncle are simple cylindric, and provided posteriorly with a few fascicles of slender bristles. The flagellum is about the length of those joints combined, and is composed of 8 articulations.

The mandibular palps (fig. 3) are of normal size, being about as long as the mandibles, and are rather densely setous, with the terminal joint somewhat curved, and narrowly truncated at the tip.

The gnathopoda in the female (figs. 4, 5) are rather strong and slightly unequal, the posterior ones (fig. 5) being the larger. The propodos in both pairs is of nearly same shape, though different in size, oval quadrangular, with the palm somewhat oblique, and defined below by an obtuse corner carrying the usual spines. In the male (see fig. 1) the gnathopoda are, as usual, more powerfully developed, though the difference between the male and female in this respect is not very pronounced.

The 2 anterior pairs of pereiopoda (fig. 6) are very robust, and somewhat resemble in structure those in *G. aralensis*. In both pairs, but especially in the 2nd, the meral joint is of very considerable size, and clothed posteriorly with numerous slender curved setæ arranged in a double row, anteriorly, with scattered fascicles of rather long bristles. The carpal joint is likewise unusually expanded, though rather short, and provided with a similar dense supply of delicate setæ. The propodal joint, on the other hand, is narrow cylindric, with a row of strong spines posteriorly, accompanied by numerous slender bristles.

The 3 posterior pairs of pereiopoda (figs. 7—9) are not nearly so slender as in *G. aralensis*, and have their outer part densely clothed with fascicles of short spines accompanied by slender bristles. The basal joint of the antepenultimate pair (fig. 7) is rounded quadrangular in form, having the inferoposteal corner produced to a rounded lobe. That of the penultimate pair (fig. 8) is only expanded in its proximal part, and is accordingly of a more triangular form. The basal joint of the last pair (fig. 9) is much larger than that of the 2 preceding pairs, forming posteriorly a broad lamellar expansion, the edge of which is evenly curved, and fringed with very small hairs. Anteriorly, as also on the inner face, this joint, like that of the 2 preceding pairs, carries several fascicles of rather long bristles.

The last pair of uropoda (fig. 10) are of moderate size, and have the basal part armed at the end below with a transverse row of 7 strong spines.

The outer ramus is about twice as long as the basal part, and rather broad, sub-foliaceous, being edged all round with slender, ciliated setæ, and having moreover 2 fascicles of spines on the outer edge. Its terminal joint is so very minute, as easily to escape attention. The inner ramus is of larger size than usual, being fully half as long as the outer, and is provided inside with 2 juxtaposed spines, followed by a regular row of slender bristles.

The telson (fig. 11) is cleft by a deep incision, somewhat widening posteriorly, into 2 rather narrow lobes, each of which carries, on the somewhat blunted tip, 3 spines.

Occurrence. — Of this form at first but a few specimens, collected by Mr. Weidemann, were sent me from the Zoological Museum. The same form was subsequently found rather plentiful in some collections made by Messrs. Andrussow and Maximowicz in the bay Karabugas. It here occurred together with *G. maeoticus* at a short distance from the shore, on a sandy bottom. A few specimens of the same form were also collected by Mr. Warpachowsky in the eastern part of the North Caspian Sea, at Stat. 63.

19. Gammarus maeoticus, Sowinsky.

(Pl. 9, figs. 12—20).

Gammarus maeoticus, Sowinsky. Les Crustacés de la mer d'Azow, p. 6, Pl. I, A, Pl. II.

Specific Characters. — Very like the preceding species, as to its outward appearance. Cephalon, however, having the lateral lobes more prominent, and narrowly rounded at the tip. Anterior pairs of coxal plates comparatively smaller and rapidly diminishing in size anteriorly; 1st pair not at all expanded distally. Last pair of epimeral plates of metasome nearly rectangular. Urosome smooth above. Eyes comparatively small, reniform. Antennæ more robust than in the preceding species, flagellum of the superior ones scarcely as long as the peduncle, accessory appendage almost half the length of the flagellum, and 5-articulate. Inferior antennæ with the 3 outer joints of the peduncle rather broad, forming posteriorly rounded expansions, densely clothed with delicate bristles. Mandibular palps exceedingly large, more than twice the length of the mandibles. Gnathopoda in female comparatively small and feeble, nearly equal-sized; in male much more powerfully developed, and rather unequal, the posterior ones being much the larger. Pereiopoda, uropoda and telson nearly agreeing in structure with those parts in *G. Weidemanni*. Length of adult male 12 mm.

Remarks. — This species was detected by Mr. Sowinsky in the Azow Sea, and the female was rather fully described by him in a Russian paper on the Crustacea of that Sea. It is very nearly allied to *G. Weidemanni*, and would seem to be generally found together with this species. On a closer examination, it may, however, easily be distinguished by the more strongly built antennæ, and especially by the very dense supply of delicate bristles on the inferior ones. Moreover, the anterior pairs of coxal plates are comparatively smaller, and the gnathopoda of the female much feebler. Finally, the mandibular palps are highly remarkable for their extraordinary development.

Description. — The length of the largest male specimens amounts to 12 mm., and this form would accordingly seem to grow to a somewhat larger size than *G. Weidemanni*.

The general form of the body (see fig. 12) closely resembles that of the said species, the back being broadly rounded and remarkably smooth.

The cephalon is somewhat produced in front between the bases of the superior antennæ, and has the lateral lobes somewhat more prominent than in the preceding species, and narrowly rounded at the tip.

The anterior pairs of coxal plates are comparatively smaller than in that species, and rather unequal, diminishing rapidly in size anteriorly. The 1st pair (see fig. 16) are scarcely broader than the 2nd, and much less deep, not being at all expanded in their outer part. The 4th pair exhibit a similar form to that in *G. Weidemanni*, but have the posterior expansion less broad.

The last 2 pairs of epimeral plates of the metasome are scarcely produced at all at the lateral corners, being nearly rectangular.

The urosome is perfectly smooth above, though, on a closer examination, 2 small, juxtaposed spinules may be found, as in the preceding species, on each side of the last segment.

The eyes are comparatively smaller than in *G. Weidemanni*, and pronouncedly reniform, having a rather deep sinus anteriorly.

The antennæ, as in that species, are rather short and subequal in length, but appear to be more strongly built. The superior ones (fig. 13) have the 1st joint of the peduncle very large and massive, considerably exceeding in length the other 2 combined. The flagellum does not quite attain the length of the peduncle, and is only composed of 8—9 articulations, each having posteriorly a rather dense fascicle of delicate bristles. The accessory appendage is nearly half as long as the flagellum, and is composed of 5 articulations having, anteriorly, coarse, spiniform bristles.

The inferior antennæ (fig. 14) have the 3 outer joints of the peduncle rather broad forming posteriorly rounded expansions, which are very densely, almost in a brushlike manner, clothed with delicate bristles. The flagellum is likewise densely setiferous, and only composed of 6 articulations.

The mandibular palps (fig. 15) are highly remarkable for their extraordinary size, being more than twice as long as the mandibles. The terminal joint is fully as long as the 2nd, both being rather broad, and densely supplied with long and curved bristles.

The gnathopoda of the female (see the figures given by Mr. Sowinsky) are comparatively poorly developed, and nearly equal-sized, thereby differing conspicuously from those in the female of *G. Weidemanni*. In the male, however, these limbs (figs. 16, 17) are much more powerful and rather unequal, the posterior ones (fig. 17) being much the stronger. The propodos of the latter is very large and broad, of oval quadrangular form, whereas in the former (fig. 16) it is rather narrow oblong.

The pereiopoda are almost exactly of same structure as in *G. Weidemanni*, and need not therefore be described in detail.

The same is also the case with the uropoda (figs. 18, 19) and the telson (fig. 20).

Occurrence. — Some specimens of this form are contained in the collection of Dr. Grimm, but without statement of locality. It was taken during the past summer in great abundance by Messrs. Andrussow and Maximoviez in the bay of Karabugas, from quite shallow depth.

Distribution. — The Azow Sea (Sowinsky).

20. Gammarus pauxillus, Grimm. MS.

(Pl. 10, figs. 1—13).

Specific Characters. — Body very slender and compressed. Cephalon with the lateral lobes somewhat obliquely truncated, anterior corner the more prominent. Coxal plates of moderate size and quite smooth. Last pair of epimeral plates of metasome slightly produced at the lateral corners. Urosome with a single spinule on each side of the last 2 segments. Eyes rather large, oblong oval. Superior antennæ much longer than the inferior, flagellum very slender, accessory appendage small, triarticulate. Gnathopoda in female comparatively feeble, and subequal in size; those in male very powerful, with the propodos in both pairs exceedingly large, and oblong oval in form. Anterior pairs of pereiopoda slender, posterior pairs moderately elongated, basal joint of last pair but slightly expanded, and scarcely larger than that of the penultimate pair. Last pair of uropoda rather elongated, outer ramus sublinear, spinous, inner small, scale-like. Telson very short, lateral lobes blunted at the tip, and without any apical spines. Lenght of adult female 4 mm., of male 6 mm.

Remarks. — This is a very small species, but easily recognizable by the slender body, the very unequal-sized antennæ, the powerful structure of the gnathopoda in the male, the comparatively small size of the basal joint of the last pair of perciopoda, and finally, the armature of the urosome.

Description. — The length of fully adult, ovigerous female specimens scarcely exceeds 4 mm.: that of male specimens is somewhat greater, attaining to 6 mm.

The body in both sexes (see figs. 1 and 14) is very slender and compressed, with the back smooth throughout.

The cephalon (see fig. 15) about equals in length the first 2 segments of the mesosome combined, and has the lateral lobes rather broad, but very obliquely truncated, so as to form in front a somewhat prominent, though blunt corner.

The anterior pairs of coxal plates are not very large, though somewhat deeper than the corresponding segments, and have their distal edge quite smooth. The 1st pair (see figs. 4 and 16) are slightly expanded in their outer part, whereas the 2 succeeding pairs (see fig. 5) are about same breadth throughout. The 4th pair (fig. 7) are somewhat broader than the preceding ones, and have the posterior expansion transversely truncated.

The 2 posterior pairs of epimeral plates of the metasome are slightly produced at the lateral corners, whereas the 1st pair, as usual, are rounded and far less deep.

The urosome (fig. 11) is smooth above, having only at the end of each segment a few very small hairs. Laterally, however, each of the last 2 segments is armed with a single well-defined spinule.

The eyes (see fig. 15) are rather large, occupying nearly the whole height of the frontal part of the head, and exhibiting an oblong oval form. The visual elements are well developed, and the pigment dark.

The superior antennæ (fig. 2) are very slender and elongated, exceeding half the length of the body, and have the 2nd joint of the peduncle about same length as the 1st, though considerably narrower. The flagellum is fully twice as long as the peduncle, and filiform, being composed of 16—20 articulations. The accessory appendage is comparatively small, and tri-articulate.

The inferior antennæ (fig. 3) are scarcely more than half as long as the superior, and of normal structure.

The gnathopoda are very different in the 2 sexes. In the female (figs. 4, 5) they are comparatively small and feeble, and nearly equal-sized, but differing somewhat in the shape of the propodos, which in the posterior pair (fig. 5) is a little more elongated, and slightly widens distally, with the palm less oblique than in the anterior ones (fig. 4). In the male, both pairs (figs. 16, 17) are very powerfully developed, with the propodos exceedingly large, especially in the posterior pair (fig. 17), and of oblong oval form, the palm being much shorter than the hind margin.

The 2 anterior pairs of pereiopoda (fig. 6) exhibit the usual slender form, and are but sparingly setous.

The 3 posterior pairs (figs. 8—10) are moderately elongated, the basal joint being in none of them particularly expanded, that of the last pair (fig. 10) not differing much either in size or shape from that of the penultimate pair (fig. 9).

The last pair of uropoda (fig. 12) are rather elongated, especially in the male (see fig. 11), projecting considerably beyond the others. The outer ramus is sub-linear in form, and exhibits, on each side, 2 or 3 fascicles of slender spines. The terminal joint of this ramus is very narrow, spiniform, and tipped with 2 unequal bristles. The inner ramus exhibits the usual rudimentary appearance.

The telson (fig. 13) is very short, being nearly twice as broad as it is long, and is cleft to the base by a rather narrow incision. The lateral lobes are obtusely truncated at the tip, and do not exhibit any trace of apical spines, whereas 2 juxtaposed, delicate bristles are found on each lobe exteriorly, at a short distance from the tip.

Occurrence. — Of this diminutive species a few, not very well preserved specimens are contained in the collection of Dr. Grimm. They were taken in the southern part of the Caspian Sea, from the considerable depth of 108 fathoms. Another specimen was found farther north, in a depth of 40 fathoms.

21. Gammarus Andrussowi, G. O. Sars, n. sp.

(Pl. 10, figs. 18—26).

Specific Characters. — Body rather slender and compressed. Lateral lobes of cephalon transversely truncated. Coxal plates comparatively larger than in *G. pauxillus*, 4th pair rather broad, and angularly produced below the posterior emargination. Epimeral plates of metasome about as in the said species. Urosome with the 2 anterior segments spinulose both dorsally and laterally. Eyes not very large, oval reniform. Superior antennae much longer than the inferior, joints of the peduncle successively diminishing in size, flagellum very slender, accessory appendage 4-articulate. Gnathopoda in female very unequal, the anterior ones being much the stronger, and having the propodos rather large; posterior ones extremely slender, with the propodos narrow, sub-linear. The 2 anterior pairs of pereiopoda normal; the 3 posterior pairs moderately elongated, with the basal joint in all of them rather expanded, and produced at the infero-posteal corner, that of last pair much the largest, and having the posterior expansion obtusely truncated below. Last pair of uropoda much elongated, with the outer ramus sub-linear and spinous, inner rudimentary. Telson of moderate size, lateral lobes obtusely pointed, and having each a rather strong lateral spine and 3 apical ones. Length of adult female 5 mm.

Remarks. — This new species, which I have much pleasure in dedicating to its discoverer, Mr. Andrussow, is prominently distinguished by the anomalous structure of the gnathopoda, which somewhat resembles that found in the genus *Iphigenella*, to be described below. Moreover, the shape of the basal joint of the 3 posterior pairs of pereiopoda, the armature of the telson, and the considerable length of the last pair of uropoda may serve for easily recognizing the species.

Description. — The length of the solitary specimen examined, which is an adult, ovigerous female, measures only 5 mm., and this form is accordingly of rather small size.

The body (see fig. 18) is very slender and much compressed, with the back narrowly rounded and quite smooth.

The cephalon is fully as long as the first 2 segments of the mesosome, and has the lateral lobes rather broad, and transversely truncated at the tip.

The anterior pairs of coxal plates are comparatively larger than in *G. pauxillus*, and, as in that species, quite smooth. The 1st pair are but very slightly expanded distally, though somewhat broader than the succeeding one. The 4th pair are rather broad, and have the posterior expansion obliquely truncated, forming below the emargination a rather projecting corner.

The epimeral plates of the metasome are nearly of same shape as in *G. pauxillus*.

The urosome (fig. 24), on the other hand, differs markedly in having the first 2 segments armed with both dorsal and lateral fascicles of spines, the dorsal fascicle containing 2—4, each of the lateral ones 3 spinules. The last segment has only a single small spinule on each side, at the base of the telson.

The eyes are much smaller than in *G. pauxillus*, and oval reniform in shape.

The superior antennae are very slender and elongated, exceeding half the length of the body, and have the joints of the peduncle successively diminishing in size (see fig. 19). The flagellum is fully twice as long as the peduncle, and is composed of about 20 articulations. The accessory appendage is somewhat longer than the last peduncular joint, and 4-articulate.

The inferior antennae are scarcely more than half as long as the superior, and of usual structure.

The gnathopoda are very unequally developed, the anterior ones (fig. 20) in this species, contrary to what is usually the case, being much the stronger. The propodos of this pair is very large and of oblong oval form, with the palm somewhat oblique, though defined below by a distinct spinous corner. The posterior gnathopoda (fig. 21) are exceedingly slender, with both the carpus and propodos rather elongated, and edged with dense fascicles of slender bristles. The propodos is about the length of the carpus, and very narrow, sublinear, and with the palm extremely short and nearly transverse.

The 2 anterior pairs of pereiopoda are rather slender, and normally developed.

The 3 posterior pairs are moderately elongated, and distinguished by the basal joint being in all of them rather broad and expanded, with the infero-posteal corner produced also in the penultimate pair (fig. 22). The basal joint of the last pair (fig. 23) is much larger than that of the 2 preceding pairs, and has the posterior expansion obtusely truncated below.

The last pair of uropoda (fig. 25) are of considerable length, being even somewhat longer than the urosome, and have the outer ramus sub-linear in form, with several fascicles of slender spines on each side. Its terminal joint is rather small and spiniform.

The telson (fig. 26) is of moderate size, the cleft somewhat widening posteriorly. The lateral lobes are obtusely pointed at the tip, and carry each a rather strong lateral spine and 3 somewhat smaller apical ones.

Occurrence. — The above-described specimen was taken last year by Mr. Andrussow, at Krasnowodsk.

Gen. **Niphargoides**, G. O. Sars.

In my 2nd article on the Caspian Amphipoda, I mentioned that a new species of this genus was contained in the collection of Dr. Grimm. This form is described below, and increases the species, as at present known, to 6 in all.

22. Niphargoides Grimmi, G. O. Sars, n. sp.

(Pl. 11, figs. 1—12).

Specific Characters. — Form of body about as in *N. caspius.* Lateral lobes of cephalon somewhat projecting and narrowly rounded. Anterior pairs of coxal plates of moderate size and only fringed thwi rather short bristles. Last pair of epimeral plates of metasome acutely produced at the lateral corners, and having a few short bristles along their posterior edge. Urosome with small, scattered spinules on the dorsal face of the last 2 segments. Eyes oval reniform. Antennæ very short and stout, equal-sized, basal joint of the superior ones exceedingly large, flagellum of the inferior one very small, 3-articulate. Gnathopoda very unequal in size, the posterior ones being nearly twice as large as the anterior, propodos in both pairs gradually widening distally, palm well defined and armed at the inferior corner with a remarkably elongated spine. The 2 anterior pairs of pereiopoda nearly as in *N. caspius;* the 3 posterior pairs somewhat more slender and less densely setous, basal joint of penultimate pair remarkably narrow, that of last pair very much expanded, with the posterior edge strongly curved above the middle, and densely fringed with very short bristles. Last pair of uropoda with the outer ramus conically tapering, and fringed with scattered setæ, inner ramus nearly half the length of the outer. Telson with the lateral lobes sub-linear, and armed each with 4 slender apical spines. Length 8 mm.

Remarks. — The present species, which was confounded by Dr. Grimm with his *N. caspius*, may, on a closer examination, be easily distinguished by the far less hirsute coxal plates, the rather different shape of the propodos of the gnathopoda, the more slender and less densely setous posterior pairs of pereiopoda, and especially by the very much expanded basal joint of the last pair.

Description. — The length of fully adult specimens does not exceed 8 mm., and this form is accordingly rather inferior in size to *N. caspius*.

The general form of the body (see fig. 1) resembles that in *N. caspius*, being rather robust and tumid, with the back broadly vaulted.

The cephalon does not nearly attain the length of the first 2 segments of the mesosome combined, and has the lateral lobes somewhat prominent and narrowly rounded at the tip.

The anterior pairs of coxal plates are somewhat deeper than the corresponding segments, and have their distal edge fringed with comparatively short bristles. The 1st pair (see fig. 4) are not expanded distally, and are nearly transversely truncated at the tip. The 4th pair (see fig. 6) are somewhat deeper than they are broad, and exhibit a distinct angular corner below the posterior emargination. The posterior pairs (see figs. 7—9) are very small.

The epimeral plates of the metasome are well developed, and rather unequal in shape, the 1st pair being rounded, the 2nd nearly rectangular, whereas the last pair are considerably produced at the lateral corners, forming an acute angle, above which a few small bristles are seen issuing from the posterior edge.

The urosome is rather stout, and has 2 very small spinules dorsally on the last 2 segments, and on the terminal segment, also a single lateral spinule.

The eyes are of moderate size, and oval reniform in shape, with dark pigment.

The superior antennæ (fig. 2) are scarcely more than twice as long as the cephalon, and but sparsely setiferous. The 1st joint of the peduncle is exceedingly large, almost equalling in length the remaining part of the antenna, whereas the last joint is extremely small, nearly as broad as it is long. The flagellum about equals in length the last 2 peduncular joints combined, and is composed of 5 articulations. The accessory appendage is about half the length of the flagellum, and biarticulate, with the terminal articulation very small.

The inferior antennæ (fig. 3) are about the length of the superior, and somewhat more densely supplied with delicate bristles, especially along the posterior edge. The antepenultimate joint of the peduncle is rather broad and expanded; the penultimate one gradually tapers distally, and projects at the end anteriorly to a setiferous corner. The last joint of the peduncle is much smaller than the penultimate, and simple cylindric in form. The flagellum is very small, being scarcely longer than the last peduncular joint, and is only composed of 3 articulations.

The gnathopoda (figs. 4, 5), as in *N. caspius*, are very unequal in size, the posterior ones being nearly twice as large as the anterior. In structure they somewhat resemble those of the said species, but differ conspicuously in the shape of the propodos. Whereas in *N. caspius*, the propodos of both pairs is conically tapered, it is in this species, on the contrary, gradually expanded distally, with the palm much less oblique, and defined below by a distinct corner carrying 2 strong spines, one of which is remarkably elongated.

The 2 anterior pairs of pereiopoda (fig. 6) are rather robust, and on the whole agree in their structure with those in *N. caspius*.

The 3 posterior pairs (figs. 7—9), on the other hand, are more slender and less densely setous, differing also conspicuously in the shape of the basal joint. In the antepenultimate pair (fig. 7) this joint is obliquely oval in form, with the anterior edge strongly curved, and the posterior one somewhat bulging in its proximal part. The basal joint of the penultimate pair (fig. 8) is much more elongated, but remarkably narrow, being scarcely at all expanded. The last pair (fig. 9) are highly remarkable for the large size of the basal joint, which forms posteriorly a very broad lamellar expansion having its greatest curvature somewhat above the middle. The edge of the expansion is throughout densely fringed with small, hair-like bristles.

The 2 anterior pairs of uropoda (fig. 10) resemble in structure those in *N. caspius*, though having a greater number of spines on the rami.

The last pair of uropoda (fig. 11) are comparatively short, and have the basal part armed at the end below with a transverse row of slender spines. The outer ramus is about twice as long as the basal part, somewhat tapering distally, and edged with a limited number of slender setæ; its terminal joint is very small, but well defined. The inner ramus is about half as long as the outer, and carries on the tip 2 spines.

The telson (fig. 12) is somewhat longer than it is broad at the base, and has the cleft rather narrow. The lateral lobes are but slightly narrowed distally, and each carry on the bluntly truncated tip, a row of 4 slender spines.

Occurrence. — This form has been collected by Dr. Grimm occasionally in 5 different Stations, belonging partly to the southern, partly to the middle part of the Caspian Sea, the depth ranging from 25 to 90 fathoms.

Gen. **Cardiophilus**, G. O. Sars, n.

Generic Characters. — Body much elongated, smooth, with the anterior pairs of coxal plates rather broad. Superior antennæ of moderate length and provided with a very small accessory appendage. Inferior antennæ extremely small. Mandibles with the cutting part divided into 2 very narrow, almost spiniform lamellæ, palp well developed. First pair of maxillæ with the palp extremely small, biarticulate. Maxillipeds with the masticatory lobes carrying, inside, flattened spines, palps with the dactylar joint rudimentary, nodiform. Gnathopoda (in female) rather feeble, the posterior ones more elongated than the anterior. Pereiopoda with strong, hooked dactyli, basal joint of the 3 posterior pairs but slightly expanded. Last pair of uropoda very small, scarcely extending beyond the others. Telson short, deeply cleft.

Remarks. — This new genus is founded upon a small Amphipod, which is stated, at least occasionally, to lead a parasitic existence, having been found within the mantle of a species of *Cardium*. The structure of the oral parts and the strongly hooked dactyli of the pereiopoda would, indeed, seem to point to a semi-parasitic habit.

23. **Cardiophilus Baeri**, G. O. Sars, n. sp.

(Pl. 11. figs. 13—27).

Specific Characters. — Body extremely slender and elongated, nearly smooth throughout. Lateral lobes of cephalon evenly rounded. Anterior pairs of coxal plates not very deep, but quadrangular in form and perfectly smooth; 4th pair but slightly emarginated posteriorly. Epimeral plates of metasome scarcely produced at the lateral corners. Urosome rather small, and perfectly smooth. Eyes of moderate size and rounded form. Superior antennæ not much elongated, peduncle rather short, with the 1st joint much the largest, flagellum exceeding the peduncle in length and composed of about 10 articulations, accessory appendage extremely minute, biarticulate. Inferior antennæ scarcely half as long as the superior. Anterior gnathopoda rather small, with the propodos oval quadrangular in form; posterior ones much more elongated, with the propodos sub-linear in shape. Pereiopoda moderately elongated and almost naked, dactylus in all of them hook-shaped, terminating in a very sharp point; basal joint of last pair not differing much from that of the 2 preceding pairs. Last pair of uropoda with the outer ramus scarcely longer than the basal part, inner scale-like. Telson semi-lunar in form, cleft very narrow. Length of adult female $5\frac{1}{2}$ mm.

Remarks. — This is as yet the only known species of the genus, and may easily be recognized from the other Caspian Amphipoda by its elongated, sub-cylindrical body and the unusually poor development of the inferior antennæ.

Description. — The length of fully adult female specimens measures only $5\frac{1}{2}$ mm., and this form is accordingly of rather small size.

The body (see fig. 13) is remarkably slender and elongated; perfectly smooth, and not at all compressed, being nearly cylindric in form. The anterior division, comprising the cephalon and mesosome, is fully twice as long as the posterior one, all its segments being of nearly equal length.

The cephalon does not exceed in length the 1st segment of the mesosome, and is but very slightly produced in front between the bases of the superior antennæ. The lateral lobes are evenly rounded, and behind them the inferior edges of the head form on each side a rather deep emargination encircling the globular basal joint of the inferior antennæ.

The anterior pairs of coxal plates are but little deeper than the corresponding segments, but comparatively broad, and they slightly increase in size posteriorly. The 1st pair (see fig. 21) are almost quadrate in form, the 2 succeeding ones oval quadrangular. The 4th pair are not much broader than the preceding pairs, and but very slightly emarginated posteriorly.

The 3 posterior pairs of coxal plates are rather small, and of usual appearance.

The epimeral plates of the metasome are not particularly large, and the last 2 are scarcely produced at all at the lateral corners.

The urosome is comparatively small and perfectly smooth, with no traces either of hairs or spines.

The eyes are well developed, and of an almost orbicular form, with dark pigment.

The superior antennæ (fig. 14) scarcely exceed in length ¼ of the body, and have the peduncle comparatively short, with the 1st joint much the largest, exceeding in length the other 2 combined. The flagellum is somewhat longer than the peduncle, and is composed of 10 very sharply-defined articulations carrying, on both edges, small, hair-like bristles. The accessory appendage is extremely small, so as easily to escape attention, the more so as it generally lies inside the base of the flagellum. On a closer examination it is, however, found to be composed of 2 well-defined articulations.

The inferior antennæ (fig. 15) are quite unusually poorly developed, being scarcely half as long as the superior, and have the basal joint globularly tumefied. The last joint of the peduncle is somewhat smaller than the penultimate one, both having at the end a few small bristles. The flagellum is about half the length of the peduncle, and is composed of 4 articulations.

The buccal area is somewhat protuberant, not being covered laterally by the 1st pair of coxal plates. The oral parts composing it, differ in some points rather markedly from those in the other Gammaridæ.

The anterior lip (not figured in the plate) is rather small and rounded, without any projection in front.

The posterior lip (fig. 17) is normally developed, with the lateral lobes slightly incurved at the tip, and produced laterally to an obtuse auricular projection.

The mandibles (fig. 16) are comparatively small, and have the cutting part divided into 2 slightly dentated lamellæ, which are remarkably narrow, almost spiniform. Between this part and the rather poorly developed molar expansion, only 3 small bristles occur in each mandible. The palp is normally developed, being considerably longer than the mandible, and is but sparsely setiferous. The terminal joint is somewhat shorter than the middle one, and somewhat compressed, with a row of small spinules along the distal part of the inner edge.

The 1st pair of maxillæ (fig. 18) are highly remarkable for the rudimentary condition of the palp, which is extremely small, so as scarcely to extend beyond the masticatory lobe. On a closer examination, it is found to be composed of 2 nearly equal-sized joints, the outer of which terminates with 2 small bristles. The masticatory lobe is rather broad and armed at the truncated tip with a number of strong, denticulated spines. The basal lobe is much smaller and carries 3 ciliated setæ.

The 2nd pair of maxillæ (fig. 19) are nearly as large as the 1st pair, and quite normal in structure.

The maxillipeds (fig. 20) have the basal and masticatory lobes of moderate size, the latter being armed inside with a row of flattened spines. The palps are but sparsely setous, and have the dactylar joint quite rudimentary, only represented by a small nodule carrying 2 minute hairs.

The gnathopoda (figs. 21, 22) are rather feeble in structure, and somewhat unequal, the posterior ones being considerably more elongated than the anterior. The latter (fig. 21) are rather short, and have the propodos comparatively small, oval quadrangular in form, and slightly narrowed distally, with the palm nearly transverse. The posterior gnathopoda (fig. 22) have both the carpus and the propodos considerably more elongated, the latter being nearly linear in form, with the palm extremely short and transverse. Both pairs are provided with scattered fascicles of slender bristles.

The pereiopoda are of moderate length, and almost naked, with only a few very small and scattered hair-like bristles. In all of them the dactylus is very strong, hook-shaped, and terminating in a very acute point. The 2 anterior pairs (fig. 23) are exactly alike, and have the meral joint somewhat expanded distally. The 3 posterior pairs successively increase somewhat in length, and have the basal joint but slightly expanded and of oval form. In the last pair (fig. 27) this joint is not very different either in shape or size from that of the penultimate pair.

The 2 anterior pairs of uropoda (fig. 25) are rather strongly built, with the rami subequal and armed at the tip with several, partly hooked spines.

The last pair of uropoda (fig. 26) are extremely small, scarcely projecting beyond the others at all. The outer ramus is scarcely longer than the basal part, and only provided with a few slender bristles; its terminal joint is so very minute, as easily to be overlooked. The inner ramus is much smaller than the outer, scale-like, and only provided with a single apical bristle.

The telson (fig. 27) is very short, being fully twice as broad as it is long, and exhibits an almost semi-lunar form, though being divided by a narrow cleft into 2 halves, each of which has a small apical, and 2 lateral hairs.

Occurrence. — Of this peculiar Amphipod, a solitary specimen was taken as early as in the year 1877 by v. Baer off the peninsula Mangyschlack. Another specimen is contained in the collection of Dr. Grimm, with the statement of having been taken at Stat. 116 within the mantle of *Cardium Baeri*. Solitary specimens were also found in the collections of Mr. Warpachowsky, taken from Stat. 16 and 31 of the North Caspian Sea.

Gen. **Iphigenella** [1]), Grimm. MS.

Generic Characters. — Body rather stout, smooth, with comparatively large coxal plates. Superior antennae longer than the inferior, and provided with a well-developed accessory appendage. Oral parts on the whole normal. Gnathopoda in both sexes very unequally developed, the anterior ones rather powerful, with the propodos very large, the posterior slender and feeble, with the propodos sub-linear. Pereiopoda comparatively short and stout, especially the 3 posterior pairs, propodal joint in all of them sub-cheliform, basal joint of the 3 posterior pairs lamellarly expanded, that of last pair somewhat differing in shape from that of the 2 preceding pairs. Last pair of uropoda not very large, outer ramus spinous, inner very small, scale-like. Telson rather narrow, and cleft to the base.

Remarks. — The most prominent feature of this genus, established by Dr. Grimm, is undoubtedly the peculiar prehensile character of the pereiopoda, which would seem to point to a semiparasitic nature of the animal. Also the structure of the gnathopoda is peculiar in the very unequal development of the 2 pairs. Besides the typical form described below, Dr. Grimm has referred another species to the same genus, under the name of *Iphigeneia abyssorum*. But the specimens of this form contained in the collection are evidently quite immature, and do not at all agree with the characters of the genus.

24. Iphigenella acanthopoda, Grimm. MS.

(Pl. 12, figs. 1—17).

Specific Characters. — Body moderately compressed, with the back evenly vaulted and smooth throughout. Lateral lobes of cephalon sub-angular in front. Anterior pairs of coxal plates much deeper than the corresponding segments; 1st pair sub-angular in front, 4th pair much deeper than they are broad. Last pair of epimeral plates of metasome acutely produced at the lateral corners. Urosome with the 2 anterior segments somewhat raised dorsally. Eyes of moderate size, oval in form. Superior antennae more slender and less densely setous than the inferior, peduncle rather short, accessory appendage 4-articulate. Anterior gnathopoda with the propodos very large, oval triangular in form, palm somewhat oblique; posterior ones with the propodos longer than the carpus, and edged with fascicles of slender bristles. The 2 anterior pairs of pereiopoda somewhat more slender than

---

[1]) As the name *Iphigeneia*, proposed by Dr. Grimm, has been long ago appropriated in Zoology, I have felt justified in changing it in the above manner.

the posterior, the latter rather robust, with the outer part spinous, basal joint of last pair obliquely expanded, that of the 2 preceding pairs regularly oval. Propodal joint in all the legs exhibiting at the end a short but well defined palm armed with several strong spines; dactylus strongly curved, unguiform. Last pair of uropoda with the outer ramus nearly 3 times as long as the basal part. Telson with the lateral lobes sub-linear, and each armed with 3 apical spines. Length of adult female 9 mm.

Remarks. — In my opinion, the above-characterized species is as yet the only one referable to this genus, its specific name being probably derived from the peculiar armature of the propodal joint of the pereiopoda.

Description. — The length of fully adult specimens measures about 9 mm.

The general form of the body (see fig. 1) appears rather robust and somewhat compressed, the back being evenly vaulted and smooth throughout.

The cephalon is comparatively short, but little exceeding in length the 1st segment of the mesosome, and forms in front, between the bases of the superior antennæ, a very small rostral projection. The lateral lobes are very obliquely truncated, so as to form in front an angular corner.

The anterior pairs of coxal plates are rather large, being considerably deeper than the corresponding segments, and are crowded together, so as to form a nearly continuous wall. The 1st pair are very slightly expanded distally, and are somewhat angular in front, their distal edge being, as in the other pairs, quite smooth. The 2 succeeding pairs are more regularly oblong quadrangular in form. The 4th pair are, as usual, the largest, being however much deeper than they are broad. They exhibit posteriorly in their upper part a distinct emargination, and are produced below the emargination to a short transversely-truncated expansion.

The 3 posterior pairs of coxal plates are somewhat more fully developed than in most other Gammaridæ, the antepenultimate pair (see fig. 13) being almost half as large as the preceding one, and having both lobes nearly equal-sized.

The epimeral plates of the metasome are of moderate size, the last pair being acutely produced at the lateral corners.

The urosome is comparatively short and stout, and slightly carinated dorsally, each of the 2 anterior segments being distinctly elevated at the hind edge. They, moreover, carry in the middle a few simple hairs, and on each side a single, well-defined spinule. The last segment is provided on each side with 2 such spinules.

The eyes are of moderate size, and oval in form, with dark pigment.

The superior antennæ (fig. 2) are rather slender, somewhat exceeding in length ⅓ of the body, and have the peduncle comparatively short, with the 1st joint much the largest. The flagellum is fully twice as long as the peduncle, and composed of about 17 articulations. The accessory appendage is well developed, and 4-articulate.

The inferior antennæ (fig. 3) are shorter than the superior, but somewhat more strongly built. The last 2 joints of the peduncle are about same length, and both provided with scattered fascicles of slender bristles. The flagellum is nearly as long as the peduncle, and composed of 10 articulations.

The oral parts (figs. 4—9) are on the whole of quite normal structure, and need not therefore be described in detail.

The gnathopoda (figs. 10, 11), on the other hand, are rather anomalous. They are very unequally developed, the anterior ones (fig. 10) being much more powerful than the posterior (fig. 11), and rather densely setiferous.

The anterior ones (fig. 10) have the carpus rather short, whereas the propodos is exceedingly large and oval triangular in form, with the palm somewhat oblique, and about the length of the hind margin. The palmar edge has in front of the middle a strong denticle, and on the lower, obtusely rounded corner 2 similar denticles are affixed. The dactylus is rather elongated and falciform.

The posterior gnathopoda (fig. 11) are extremely slender, and also considerably more elongated than the anterior, with the carpus much longer. The propodos somewhat exceeds the carpus in length, and is very narrow, sub-linear in form, with the palm extremely short and transverse. It is, like the carpus, provided with numerous fascicles of rather elongated bristles, which are more crowded on the lower edge.

The 2 anterior pairs of pereiopoda (fig 12) are exactly alike in structure, and rather slender, being edged with scattered fascicles of delicate bristles.

The 3 posterior pairs, unlike what is generally the case, are shorter than the anterior, being nearly of uniform length. They are rather robust, and have their outer part armed with fascicles of strong spines. The basal joint is about same size in all 3 pairs, though differing somewhat in shape, that of the 2 anterior ones (fig. 13) being oval quadrangular and somewhat broader in its proximal part, whereas in the last pair (fig. 14) this joint is obliquely expanded, so as to form at the infero-posteal corner a rather projecting rounded lobe.

In all the pereiopoda the propodal joint exhibits a rather anomalous structure (see fig. 14a), being somewhat compressed, and gradually widening towards the end, where it forms a short, but well-defined palm, armed with a number of strong denticles, against which the strongly curved, unguiform dactylus admits of being impinged, thus constituting an imperfect chela.

The 2 anterior pairs of uropoda (fig. 15) have the outer ramus considerably shorter than the inner, both being linear in form, and tipped by numerous strong spines.

The last pair of uropoda (fig. 16) are moderately elongated, projecting considerably beyond the others. The outer ramus is somewhat flattened, though not very broad, and gradually tapers distally. It is provided on the outer edge with 2, on the inner with 3 fascicles of spines accompanied by a few slender bristles, and has the terminal joint very narrow, and spiniform. The inner ramus is extremely small, and scale-like.

The telson (fig. 17) is comparatively narrow, being almost twice as long as it is broad, and scarcely tapering at all distally. It is divided by a deep and narrow cleft into 2 halves, each of which carries, at the obtusely rounded tip, 3 short spinules.

Occurrence. — Of this form, some specimens are contained in the collection of Dr. Grimm, having been taken as early as in the year 1871 by Prof. Kessler, at Astrachan, from *Astacus leptodactylus*, and erroneously labelled *Gammarus pulex*. Some other specimens, for the most part not yet fully grown, were collected by Dr. Grimm in the bay of Baku. Moreover a single small specimen was taken last summer by Mr. Maximowicz, in the bay of Karabugas.

Fam. COROPHIIDÆ.

Gen. **Corophium**, Latr.

Of this genus, which has previously been regarded as exclusively marine, no less than 6 different Caspian species have been described by the author in his 3rd article on the Amphipoda. A 7th species is now added, easily distinguishable from all the others.

25. **Corophium spinulosum**, G. O. Sars, n. sp.

(Pl. 12, figs. 18—25).

Specific Characters. — Body of the usual depressed form. Frontal edge of cephalon not produced in the middle, lateral lobes narrowly rounded. First pair of coxal plates with 4 slender apical bristles, and several smaller ones anteriorly. Segments of urosome all well defined, and armed dorsally, near the posterior edge, with a transverse series of slender spines. Superior antennæ rather slender and elongated, 1st joint of the peduncle about same length as the 2nd, and armed below with 2 distant spines, flagellum longer than the peduncle. Inferior antennæ, as usual, much stronger in male than in female, penultimate joint of the peduncle produced at the end to a long, thumb-like projection having a very small secondary denticle inside near the base, the projection, being in female rather narrow, spiniform, in male much broader, lanceolate, and extending to the end of the last peduncular joint; the latter without any lateral protuberance, but produced at the end to a somewhat blunted projection, which, however, in female is quite rudimentary. Posterior gnathopoda with the dactylus not denticulate inside. Anterior pairs of pereiopoda rather slender, with the meral joint but slightly expanded. Last pair of pereiopoda with the outer joints narrow and partly edged with strong spines. The 2 anterior pairs of uropoda very densely spinous. Last pair of uropoda with the outer joint carrying several spines in addition to the setæ. Telson with an erect spine on each side of the base. Length of adult female $9\frac{1}{2}$ mm.

Remarks. — The present new species is prominently distinguishable by the spinous armature of the urosome, not found in any of the other species, and also some of the appendages, which are generally without spines, are in this species provided with such, for instance the last pair of pereiopoda, the telson and the last pair of uropoda. Its nearest ally would seem to be *C. chelicorne*, but it also differs in several other points from this species, for instance in the much more elongated superior antennæ, and the likewise more slender anterior pairs of pereiopoda, as also in the fact of the dactylus of the posterior gnathopoda not being denticulated.

Description. — The length of a fully adult, ovigerous female measures about $9\frac{1}{2}$ mm. and this form accordingly grows to a somewhat larger size than *C. chelicorne*.

The general form of the body (see fig. 18) is that characteristic of the genus, though perhaps somewhat more slender than in *C. chelicorne*.

The cephalon about equals in length the first 2 segments of the mesosome combined, and has the frontal margin not at all produced in the middle, but only slightly arcuate. The lateral lobes are not very prominent, and are narrowly rounded at the tip.

The 1st pair of coxal plates exhibit the usual triangular form, and carry on the tip 4 slender setæ, and along the anterior edge a number of much smaller bristles.

The epimeral plates of the metasome, as in the other species, are extended laterally, and densely fringed with bristles.

The urosome (fig. 22) has all its 3 segments very distinctly defined, and armed dorsally, at some distance from the posterior edge, with a transverse row of spines, their number on the 2 anterior segments being from 6 to 10, on the last, only 2. Besides these, each of the 2 anterior segments is armed laterally, at the insertion of the uropoda, with a number of similar spines, somewhat irregularly arranged.

The eyes are very small and rounded, being placed, as usual, at the bases of the lateral lobes.

The superior antennæ are of about same structure in the two sexes, being rather slender and but sparsely setiferous. In the female they nearly attain half the length of the body, and have the 1st joint of the peduncle about as long as the 2nd, and armed below with 2 distant spines. The flagellum somewhat exceeds the peduncle in length, and is composed of about 14 articulations.

The inferior antennæ are, as usual, somewhat different in the two sexes. In the female (see figs. 18, 19) they are of moderate size, with the penultimate joint of the peduncle but slightly widening distally, and produced at the end to a very acute spiniform process reaching about to the middle of the last peduncular joint. The latter is somewhat shorter than the penultimate one, and simple cylindric, without any trace of a lateral denticle, but with a very slight, nodiform prominence at the end inside. The flagellum does not attain to quite the length of the last peduncular joint, but exhibits the usual structure.

In the male these antennæ (fig. 26) are much more strongly developed, and are especially distinguished by the large size of the thumb-like process issuing from the penultimate joint of the peduncle. This process is rather broad at the base, of an almost lanceolate form, and extends as far as the end of the last peduncular joint. It has, moreover, inside, near the base, a distinct, though rather small, dentiform projection, of which only a very slight rudiment is found in the female. The last joint of the peduncle is produced at the end inside to a well-defined conical projection, which is crossed by the tip of the thumb-like process of the preceding joint, when bent in against it, whereby these antennæ acquire a complete cheliform character, as is the case in both sexes of *C. chelicorne*.

The gnathopoda resemble in structure those in the other species, with this difference, however, that the posterior ones (fig. 20) have the dactylus smooth, with only a few slender hairs, whereas in all the other Caspian species, it is coarsely denticulate inside.

The 2 anterior pairs of pereiopoda (fig. 21) are considerably more slender than in *C. chelicorne*, more resembling in this respect those in *C. nobile*, the meral joint being but very slightly expanded. This joint does not, however, in either of the two sexes exhibit that dense supply of setæ characteristic of the last-named species. The dactylus is extremely slender and elongated, even exceeding somewhat the propodal joint in length.

The last pair of pereiopoda (fig. 22) have the basal joint expanded in the usual manner and densely clothed with long ciliated setæ on both edges. The outer joints are rather slender and are partly armed with short spines in addition to the usual bristles.

The 2 anterior pairs of uropoda (see figs. 22, 23) are very densely supplied with spines both on the basal part and the rami, their number being considerably greater than in *C. chelicorne*.

The last pair of uropoda (fig. 24) are likewise distinguished by the presence of a number of spines on the lamellar terminal joint, which also carries the usual slender setæ.

The telson (fig. 25) exhibits a shape similar to that in the other species, but is prominently distinguishable by the presence of a rather conspicuous erect spine on each side of its base, not found in any of the other species.

Occurrence. — A few specimens of this form were collected by Dr. Grimm in the southern part of the Caspian Sea, from a depth of 25 fathoms.

List of Caspian Amphipoda

up to the present more closely examined.

(With statements of additional localities).

1. *Pseudalibrotus caspius*, Grm.
2.   "   *platyceras*, Grm.
3. *Pontoporeia microphthalma*, Grm.
4. *Boeckia spinosa*, Grm. St. 65, 67 (Warpachowsky).
5. *Gmelina costata*, Grm. St. 78 (Warp.), sin. Karabugas.
6.   "   *Kusnezowi*, Sowinsky. St. 93 (Warp.).
7.   "   *læviuscula*, G. O. Sars.
8.   "   *pusilla*, G. O. S.
9. *Gmelinopsis tuberculata*, G. O. S.
10.   "   *aurita*, G. O. S.
11. *Gammaracanthus caspius*, Grm.
12. *Amathillina cristata*, Grm. St. 63, 69, 78, 83 (Warp.); off the mouth of Terek (Kusnezow).
13.   "   *spinosa*, Grm.
14.   "   *affinis*, G. O. S. Sin. Karabugas.
15.   "   *Maximowiczi*, G. O. S.
16.   "   *pusilla*, G. O. S.
17. *Gammarus caspius*, Eichw. St. 78 (Warp.). Krasnowodsk, sin. Karabugas.
18.   "   *hæmobaphes*, Eichw. St. 65, 67, 69, 75, 83, 86 (Warp.), Krasnowodsk, sin. Karabugas, off the mouth of Terek.
19.   "   *Grimmi*, G. O. S.
20.   "   *subnudus*, G. O. S.
21.   "   *macrocephalus*, Grm.
22.   "   *tenellus*, G. O. S.
23.   "   *placidus*, Grm.

24. *Gammarus Warpachowskyi*, G. O. S. St. 78 (Warp.), sin. Karabugas.
25. » *pauxillus*, Grm.
26. » *Andrussowi*, G. O. S.
27. » *minutus*, G. O. S. sin. Karabugas.
28. » *aralensis*, Uljj., var. *caspia*, St. 78 (Warp.), sin. Karabugas.
29. » *macrurus*, G. O. S.
30. » *compressus*, G. O. S. St. 75 (Warp.).
31. » *similis*, G. O. S.
32. » *crassus*, Grm. St. 78, 83 (Warp.), sin. Karabugas.
33. » *abbreviatus*, G. O. S. St. 75 (Warp.).
34. » *obesus*, G. O. S. St. 86, 90 (Warp.), sin. Karabugas.
35. » *Weidemanni*, G. O. S.
36. » *maeoticus*, Sow.
37. » *platycheir*, G. O. S.
38. *Niphargoides caspius*, Grm. St. 70 (Warp.), sin. Karabugas.
39. » *Grimmi*, G. O. S.
40. » *corpulentus*, G. O. S.
41. » *compactus*, G. O. S.
42. » *quadrimanus*, G. O. S
43. » *aequimanus*, G. O. S.
44. *Pandorites podoceroides*, Grm.
45. *Iphigenella acanthopoda*, Grm.
46. *Cardiophilus Baeri*, G. O. S.
47. *Corophium nobile*, G. O. S.
48. » *chelicorne*, G. O. S.
49. » *curvispinum*, G. O. S. Krasnowodsk; sin. Karabugas.
50. » *robustum*, G. O. S. St. 63 (Warp.); sin. Karabugas.
51. » *mucronatum*, G. O. S. St. 59, 64, 72 (Warp.); sin. Karabugas.
52. » *monodon*, G. O. S.
53. » *spinulosum*, G. O. S.

Explanation of the plates.

Pl. 1.

Pseudalibrotus caspius, Grimm.

Fig. 1. Female, seen from left side.
» 2. Superior antenna.
» 3. Inferior antenna.
» 4. Cephalon with the buccal mass, but without the antennæ, viewed from left side.
» 5. Anterior lip from left side.
» 6. Posterior lip.
» 7. Right mandible with palp.
» 8. First maxilla.
» 9. Second maxilla.
» 10. Maxillipeds (right palp omitted).
» 11. Anterior gnathopod with coxal plate.
» 11a. Propodos of same, more highly magnified.
Fig. 12. Posterior gnathopod, without the coxal plate.
» 12a. Propodos of same, more highly magnified.
» 13. First pereiopod, with coxal plate.
» 14. Base of 2nd pereiopod, with coxal plate.
» 15. Antepenultimate pereiopod.
» 16. Penultimate pereiopod (the 2 outer joints omitted).
» 17. Last pereiopod.
» 18. Second uropod.
» 19. Last uropod.
» 20. Telson.

Pseudalibrotus platyceras, Grimm.

Fig. 21. Adult female, seen from left side.
» 22. Proximal part of superior antenna.
Fig. 23. Propodos of anterior gnathopod.

Pl. 2.

Pontoporeia microphthalma, Grimm.

Fig. 1. Female seen from left side.
" 2. Outer part of superior antenna.
" 3. Frontal part of cephalon, lateral view.
" 4. Dorsal part of urosome, with telson and last uropod, lateral view.

Fig. 5. Anterior gnathopod (basal joint not fully drawn).
" 6. Outer part of posterior gnathopod.
" 7. Last pereiopod.

Gmelina læviuscula, G. O. Sars.

Fig. 8. Adult male, seen from right side.
" 9. Part of superior antenna, comprising extremity of peduncle, accessory appendage, and base of flagellum.

Fig. 10. Anterior part of cephalon, lateral view.
" 11. Last uropod.
" 12. Telson.

Gmelina pusilla, G. O. Sars.

Fig. 13. Adult female, seen from left side.
" 14. Cephalon with bases of the antennæ, lateral view.
" 15. Superior antenna.
" 16. Anterior gnathopod with coxal plate.
" 17. Posterior gnathopod, with coxal plate, incubatory and branchial lamellæ.

Fig. 18. Last pereiopod.
" 19. Second uropod.
" 20. Last uropod.
" 21. Telson.

Pl. 3.

Gmelinopsis tuberculata, G. O. Sars.

Fig. 1. Adult female, seen from left side.
" 2. Cephalon, without antennæ and oral parts, lateral view.
" 3. Superior antenna.
" 3a. Accessory appendage, more highly magnified.
" 4. Inferior antenna.
" 5. Anterior lip.
" 6. Posterior lip.
" 7. Left mandible with palp.
" 8. First maxilla.
" 9. Second maxilla.

Fig. 10. Maxillipeds, without the left palp.
" 11. Anterior gnathopod with coxal plate.
" 12. Posterior gnathopod with coxal plate, incubatory and branchial lamellæ.
" 13. First pereiopod.
" 14. Antepenultimate pereiopod (outer part not drawn).
" 15. Penultimate pereiopod.
" 16. Last pereiopod (outer part not drawn).
" 17. Second uropod.
" 18. Last uropod.
" 19. Telson.

Gmelinopsis aurita, G. O. Sars.

Fig. 20. Female, seen from right side.
" 21. Cephalon, seen from above.
" 22. Same with the antennæ, lateral view.

Fig. 23. Extremity of last segment of urosome, with telson; dorsal view.

Pl. 4.

Gammaracanthus caspius, Grimm.

Fig. 1. Adult female, seen from left side.
" 2. Part of superior antenna, comprising the last 2 joints of the peduncle, the accessory appendage, and the base of the flagellum.

Fig. 3. Anterior gnathopod.
" 4. Posterior gnathopod (basal joint not fully drawn).
" 5. Extremity of 1st pereiopod.
" 6. Extremity of penultimate pereiopod.

Amathillina spinosa, Grimm.

Fig. 7. Adult male, seen from right side.
" 8. Part of superior antenna, exhibiting the accessory appendage and the base of the flagellum.
" 9. Anterior gnathopod.
" 10. Posterior gnathopod (the basal joint not fully drawn).
" 11. Anterior gnathopod of female, with coxal plate.

Fig. 12. Posterior gnathopod of same.
" 13. Penultimate pereiopod of female (the outer joints not drawn).
" 14. Last pereiopod of same.
" 15. Last uropod.
" 16. Telson.

Pl. 5.

Amathillina Maximowiczi, G. O. Sars.

Fig. 1. Adult female, seen from left side.
» 2. Superior antenna.
» 3. Part of same, more highly magnified, exhibiting the accessory appendage and the base of the flagellum.
» 4. Anterior gnathopod, with coxal plate.
» 5. Posterior gnathopod, with coxal plate, branchial and incubatory lamellæ.
» 6. Coxal plate of 4th pair.
» 7. Antepenultimate pereiopod.
» 8. Penultimate pereiopod.

Fig. 9. Last pereiopod.
» 10. Anterior gnathopod of male.
» 11. Posterior gnathopod of same.
» » Proximal part of penultimate pereiopod of male.
» 12. Last pereiopod of same (the outer joints omitted).
» 13. Last uropod.
» 14. Extremity of last segment with the telson, dorsal view.

Amathillina pusilla, G. O. Sars.

Fig. 15. Adult female, seen from left side.
» 16. Accessory appendage of superior antenna.
» 17. Same of a male specimen.
» 18. Anterior gnathopod of female.
» 19. Posterior gnathopod of same (basal joint not fully drawn).
» 20. First pereiopod.
» 21. Antepenultimate pereiopod (extremity omitted).

Fig. 22. Penultimate pereiopod.
» 23. Last pereiopod.
» 24. Last uropod.
» 25. Telson.
» 26. Anterior gnathopod of male.
» 27. Posterior gnathopod of same (basal joint not fully drawn).
» 28. Proximal part of last pereiopod of a male specimen.

Pl. 6.

Gammarus Grimmi, G. O. Sars.

Fig. 1. Adult female, seen from left side.
» 2. Part of superior antenna, exhibiting the accessory appendage and the base of the flagellum.
» 3. Dorsal part of urosome, lateral view.
» 4. Outer part of anterior male gnathopod.
» 5. Posterior gnathopod of male.
» 6. Anterior gnathopod of female, with coxal plate.

Fig. 7. Posterior gnathopod of same, with coxal plate, branchial and incubatory lamellæ.
» 8. Last pereiopod.
» 9. Last uropod.
» 10. Telson.
» 10a. Extremity of right lateral lobe of same, more highly magnified.

Gammarus submudus, G. O. Sars.

Fig. 11. Adult female, seen from left side.
» 12. Part of superior antenna, with the accessory appendage and the base of the flagellum.
» 13. Anterior gnathopod.
» 14. Posterior gnathopod.

Fig. 15. Proximal part of antepenultimate pereiopod.
» 16. Penultimate pereiopod.
» 17. Last pereiopod.
» 18. Last uropod.
» 19. Telson.

Pl. 7.

Gammarus macrocephalus, Grimm.

Fig. 1. Adult male, seen from left side.
» 2. Part of superior antenna, comprising the last 2 joints of the peduncle, the accessory appendage, and the base of the flagellum.
» 3. Anterior gnathopod.
» 4. Posterior gnathopod.
» 5. First pereiopod.

Fig. 6. Proximal part of antepenultimate pereiopod.
» 7. Same part of penultimate pereiopod.
» 8. Last pereiopod, without the outer joints.
» 9. Last uropod.
» 10. Second uropod.
» 11. Telson.

Gammarus tenellus, G. O. Sars.

Fig. 12. Adult female, seen from right side.
» 13. Part of superior antenna, with the accessory appendage and the base of the flagellum.
» 14. Anterior gnathopod.
» 15. Posterior gnathopod, with coxal plate, branchial and incubatory lamellæ.
» 16. Second pereiopod with coxal plate.
» 17. Antepenultimate pereiopod.

Fig. 18. Penultimate pereiopod, without the outer joints.
» 19. Last pereiopod.
» 20. Last uropod.
» 21. Second uropod.
» 22. Telson.
» 23. Anterior gnathopod of male, with coxal plate.
» 24. Posterior gnathopod of same.

Pl. 8.

Gammarus placidus, Grimm.

Fig. 1. Adult female, seen from left side.
» 2. Cephalon, without antennæ and oral parts; lateral view.
» 3. Part of superior antenna.
» 4. Anterior gnathopod, with coxal plate.
» 5. Posterior gnathopod, with coxal plate, branchial and incubatory lamellæ.
» 6. Second pereiopod, with coxal plate, branchial and incubatory lamellæ.

Fig. 7. Proximal part of antepenultimate pereiopod.
» 8. Penultimate pereiopod, without the outer joints.
» 9. Last pereiopod.
» 10. Second uropod.
» 11. Last uropod.
» 12. Telson.

Gammarus platycheir, G. O. Sars.

Fig. 14. Adult male, seen from left side.
» 15. Outer part of superior antenna.
» 16. Posterior gnathopod.

Fig. 17. Extremity of urosome, with left last uropod and telson; dorsal view.

Pl. 9.

Gammarus Weidemanni, G. O. Sars.

Fig. 1. Adult male, seen from left side.
» 2. Superior antenna.
» 3. Inferior antenna.
» 3 (bis). Mandibular palp.
» 4. Anterior gnathopod of female, with coxal plate.
» 5. Posterior gnathopod of same, with coxal plate, branchial and incubatory lamellæ.

Fig. 6. Second pereiopod.
» 7. Antepenultimate pereiopod.
» 8. Penultimate pereiopod.
» 9. Last pereiopod.
» 10. Last uropod.
» 11. Telson.

Gammarus maeoticus, Sowinsky.

Fig. 12. Adult male, seen from right side.
» 13. Superior antenna.
» 14. Inferior antenna.
» 15. Mandible with palp.
» 16. Anterior gnathopod of male, with coxal plate.

Fig. 17. Posterior gnathopod of same.
» 18. Second uropod.
» 19. Last uropod.
» 20. Telson.

Pl. 10.

Gammarus pauxillus, Grimm.

Fig. 1. Adult female, seen from left side.
» 2. Superior antenna.
» 3. Inferior antenna.
» 4. Anterior gnathopod, with coxal plate.
» 5. Posterior gnathopod, with coxal plate, branchial and incubatory lamellæ.
» 6. First pereiopod.
» 7. Coxal plate of 4th pair.
» 8. Antepenultimate pereiopod (outer part not drawn).
» 9. Penultimate pereiopod, without the outer joints.

Fig. 10. Last pereiopod.
» 11. Urosome of male, seen from left side.
» 12. Last uropod of female.
» 13. Telson.
» 14. Adult male, seen from left side.
» 15. Cephalon of same, without antennæ and oral parts; lateral view.
» 16. Anterior gnathopod of same, with coxal plate.
» 17. Posterior gnathopod.

Gammarus Andrussowi, G. O. Sars.

Fig. 18. Adult female, seen from left side.
» 19. Superior antenna (extremity of flagellum not drawn).
» 20. Anterior gnathopod (basal joint not fully drawn).
» 21. Posterior gnathopod (do.).

Fig. 22. Proximal part of penultimate pereiopod.
» 23. Last pereiopod.
» 24. Urosome, seen from right side.
» 25. Last uropod.
» 26. Telson.

Pl. 11.

Niphargoides Grimmi, G. O. Sars.

Fig. 1. Adult male, seen from left side.
» 2. Superior antenna.
» 3. Inferior antenna.
» 4. Anterior gnathopod, with coxal plate.
» 5. Posterior gnathopod.
» 6. Second pereiopod, with coxal plate.

Fig. 7. Antepenultimate pereiopod.
» 8. Penultimate pereiopod.
» 9. Last pereiopod.
» 10. Second uropod.
» 11. Last uropod.
» 12. Telson.

Cardiophilus Baeri, G. O. Sars.

Fig. 13. Adult female, seen from left side.
» 14. Superior antenna.
» 15. Inferior antenna.
» 16. Right mandible with palp, and masticatory part of left one.
» 17. Posterior lip.
» 18. First maxilla.
» 19. Second maxilla.
» 20. Maxillipeds.

Fig. 21. Anterior gnathopod, with coxal plate.
» 22. Posterior gnathopod, with part of coxal plate.
» 23. First pereiopod.
» 24. Last pereiopod.
» 25. Second uropod.
» 26. Last uropod.
» 27. Telson.

Pl. 12.

Iphigenella acanthopoda, Grimm.

Fig. 1. Adult female, seen from left side.
» 2. Superior antenna.
» 3. Inferior antenna.
» 4. Anterior lip.
» 5. Posterior lip.
» 6. Left mandible with palp.
» 7. First maxilla.
» 8. Second maxilla.
» 9. Maxillipeds (left palp omitted).
» 10. Anterior gnathopod.

Fig. 11. Posterior gnathopod (basal part not fully drawn).
» 12. First pereiopod.
» 13. Antepenultimate pereiopod.
» 14. Last pereiopod.
» 14a. Propodal joint and dactylus of same, more highly magnified.
» 15. Second uropod.
» 16. Last uropod.
» 17. Telson.

Corophium spinulosum, G. O. Sars.

Fig. 18. Adult female, seen from right side.
» 19. Part of inferior antenna of same, comprising the last 2 peduncular joints.
» 20. Posterior gnathopod.
» 20a. Dactylus of same, more highly magnified.
» 21. First pereiopod.

Fig. 22. Last pereiopod.
» 22 (bis). Urosome, seen from above (the 2 anterior uropoda on left side omitted).
» 23. Second uropod.
» 24. Last uropod.
» 25. Telson.
» 26. Inferior antenna of male.

Amphipoda Suppl   G.O.Sars Crustacea caspia.   Pl. I.

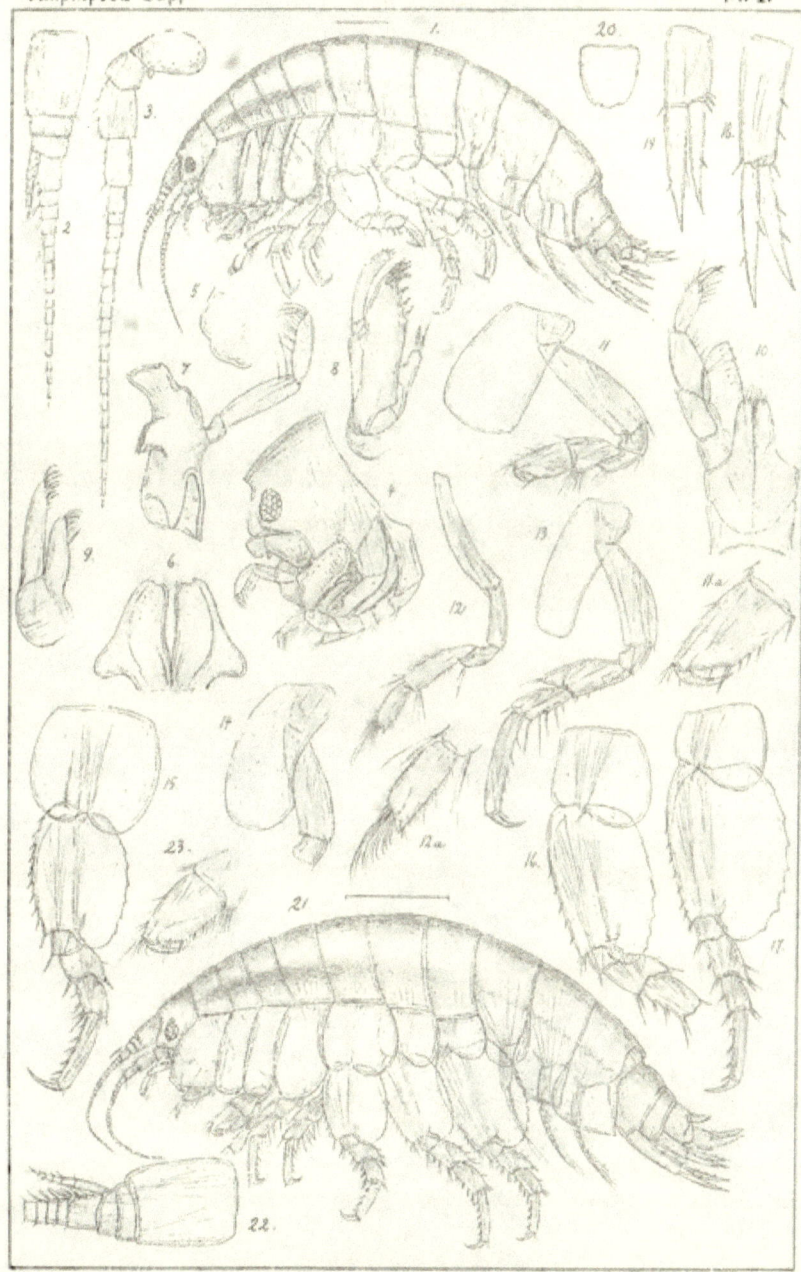

G.O.Sars autogr.   Pseudalibrotus caspius, (Grimm).
                 "   platyceras, (Grimm).

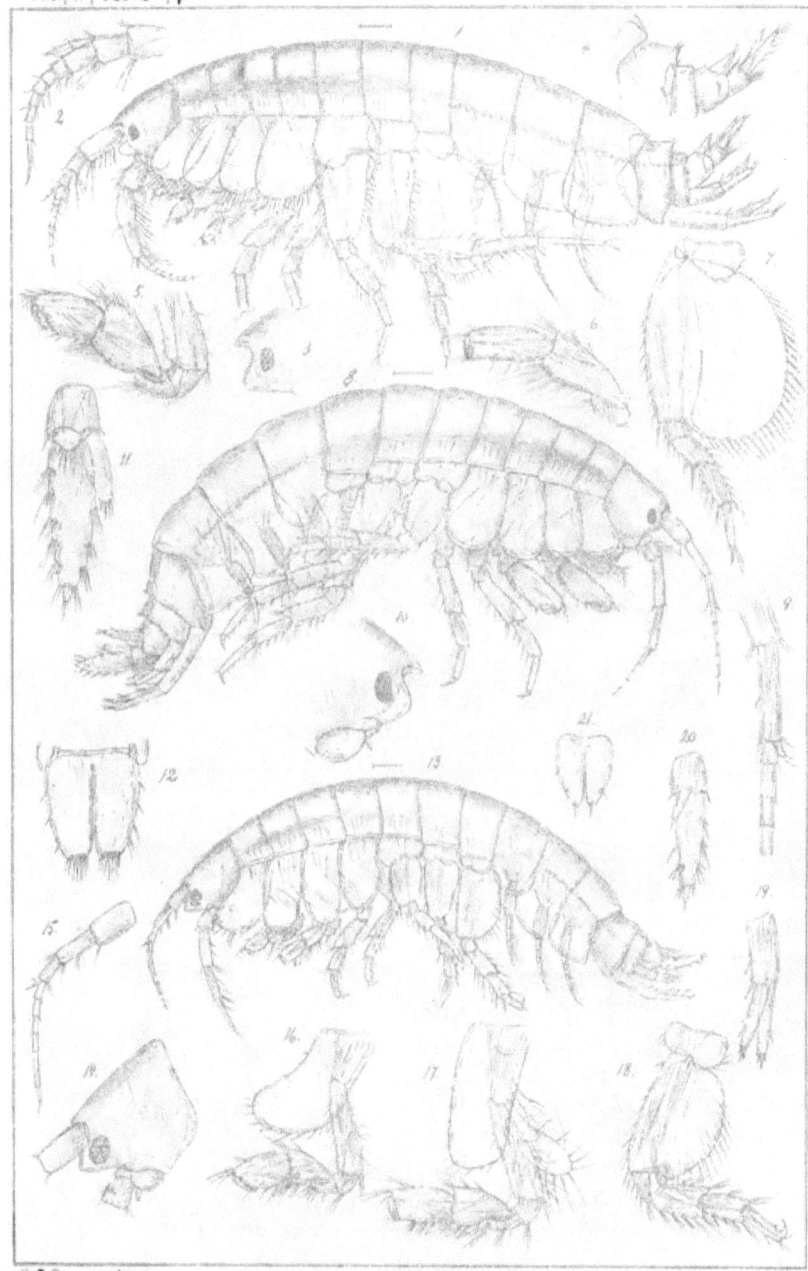

Pontoporeia microphthalma, Grimm.
Gmelina læviuscula, G.O.Sars.
pusilla, G.O.Sars.

Amphipoda Suppl. G.O.Sars Crustacea caspia. Pl. 3.

Gmelinopsis tuberculata, G.O.Sars.
" aurita, G.O.Sars.

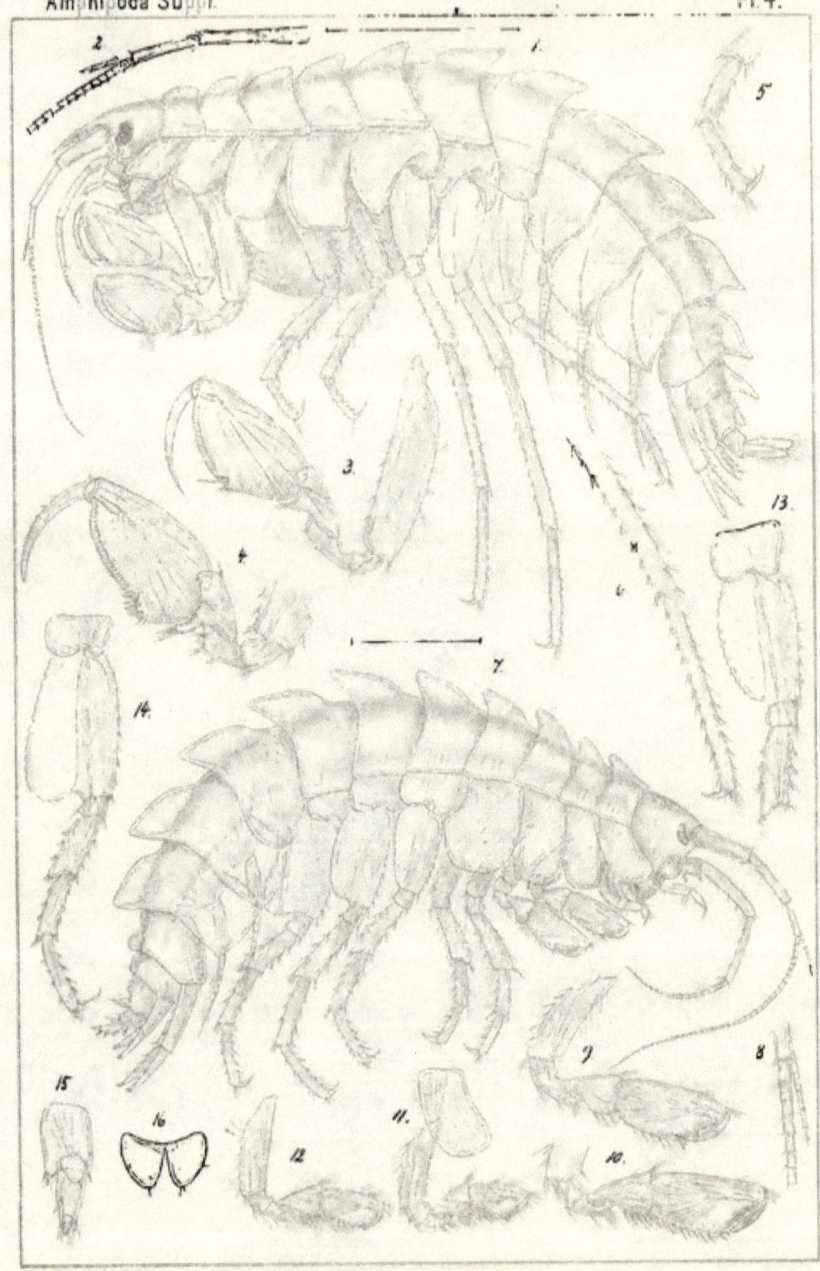

Gammaracanthus caspius, Grimm.
Amathillina spinosa, (Grimm).

Amphipoda Suppl. G.O.Sars Crustacea caspia. Pl. 5.

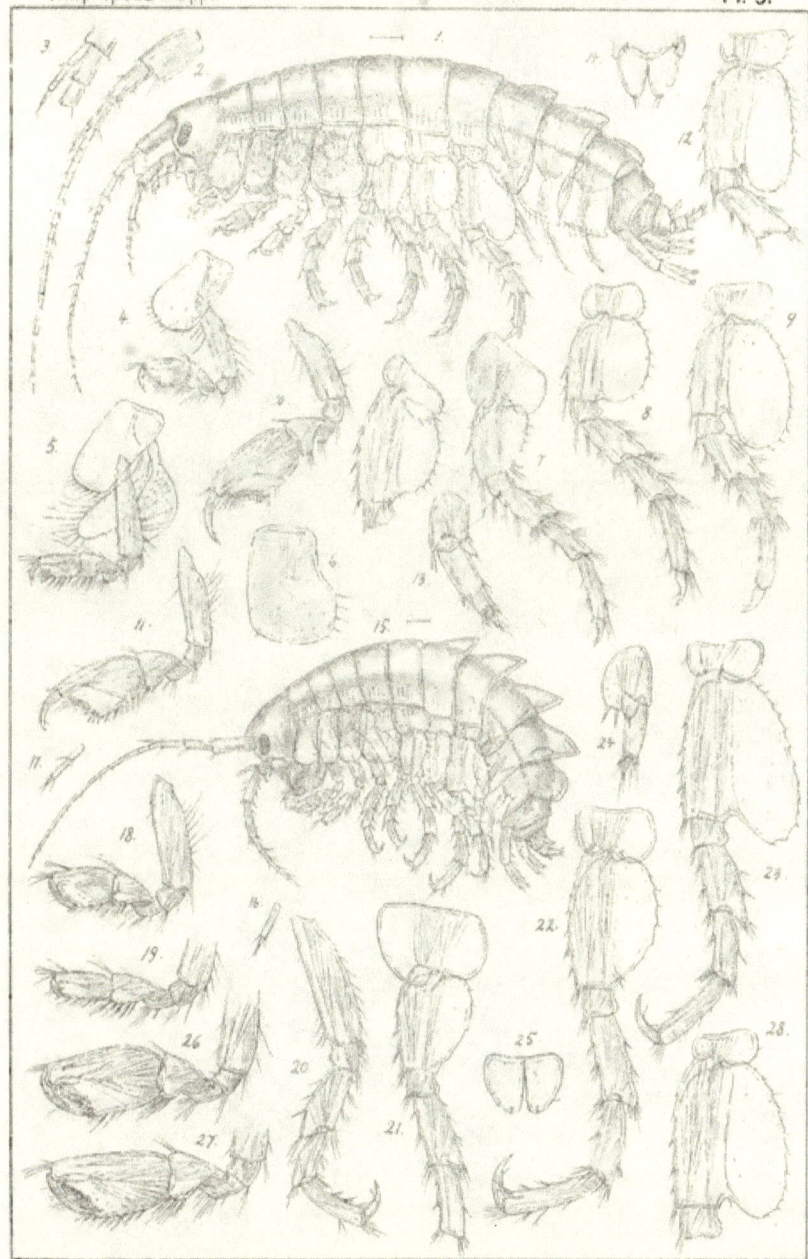

G.O Sars autogr. Amathillina Maximovitschi, G.O. Sars.
" pusilla, G.O.Sars.

Gammarus Grimmi, G.O.Sars.
subnudus G.O.Sars.

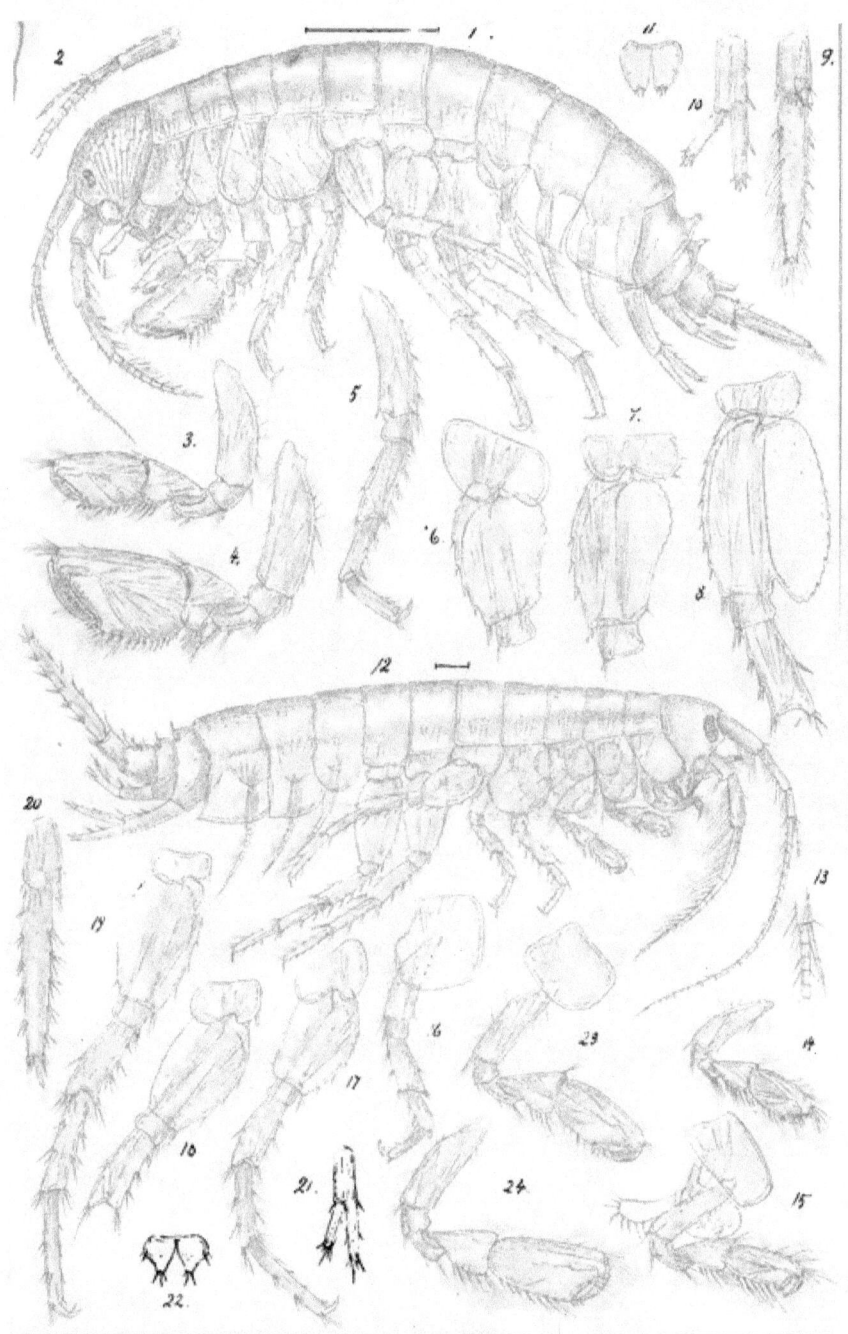

Gammarus macrocephalus, Grimm.
" tenellus, G.O.Sars.

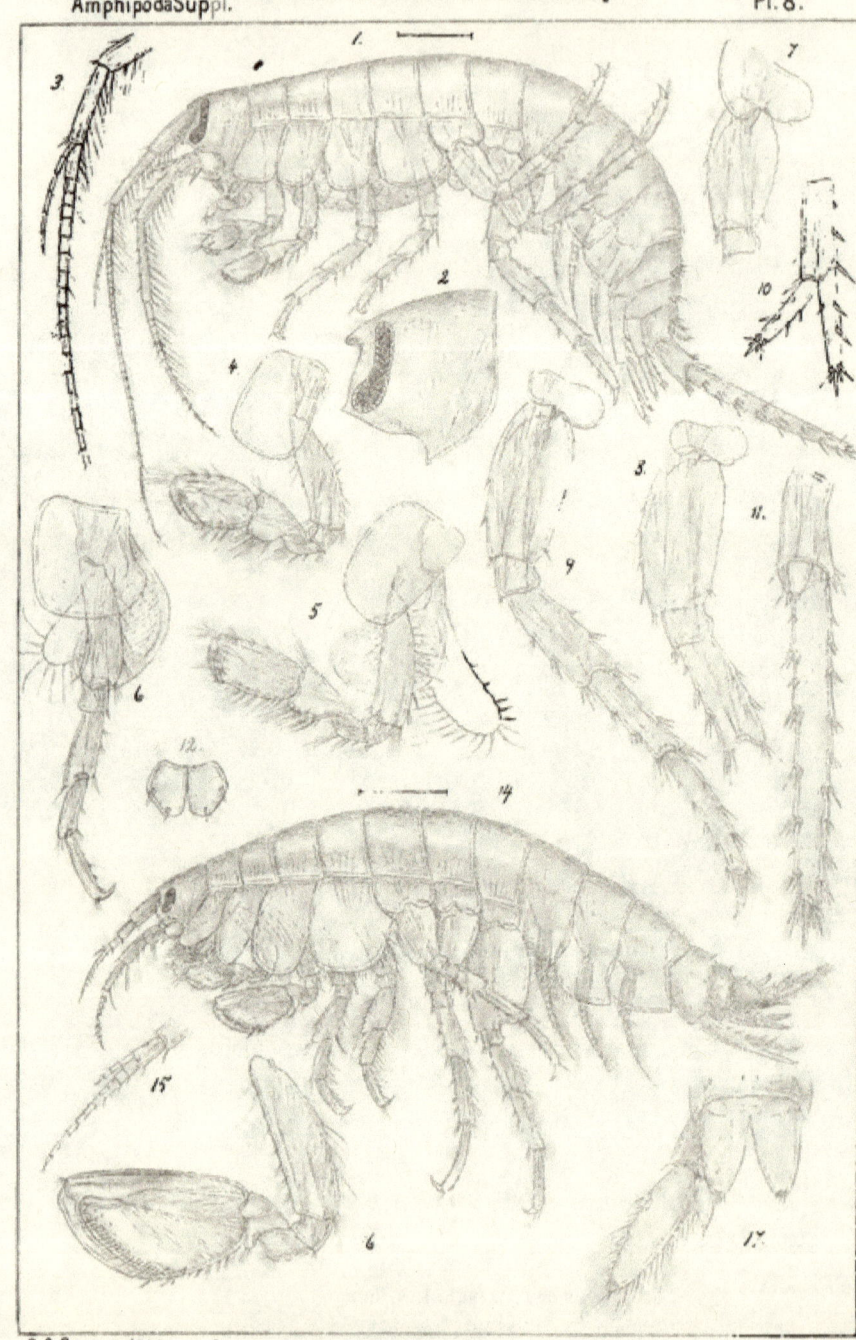

Gammarus placidus, Grimm.
" platycheir, G. O. Sars.

Amphipoda Suppl.   G.O.Sars Crustacea caspia.   Pl.9.

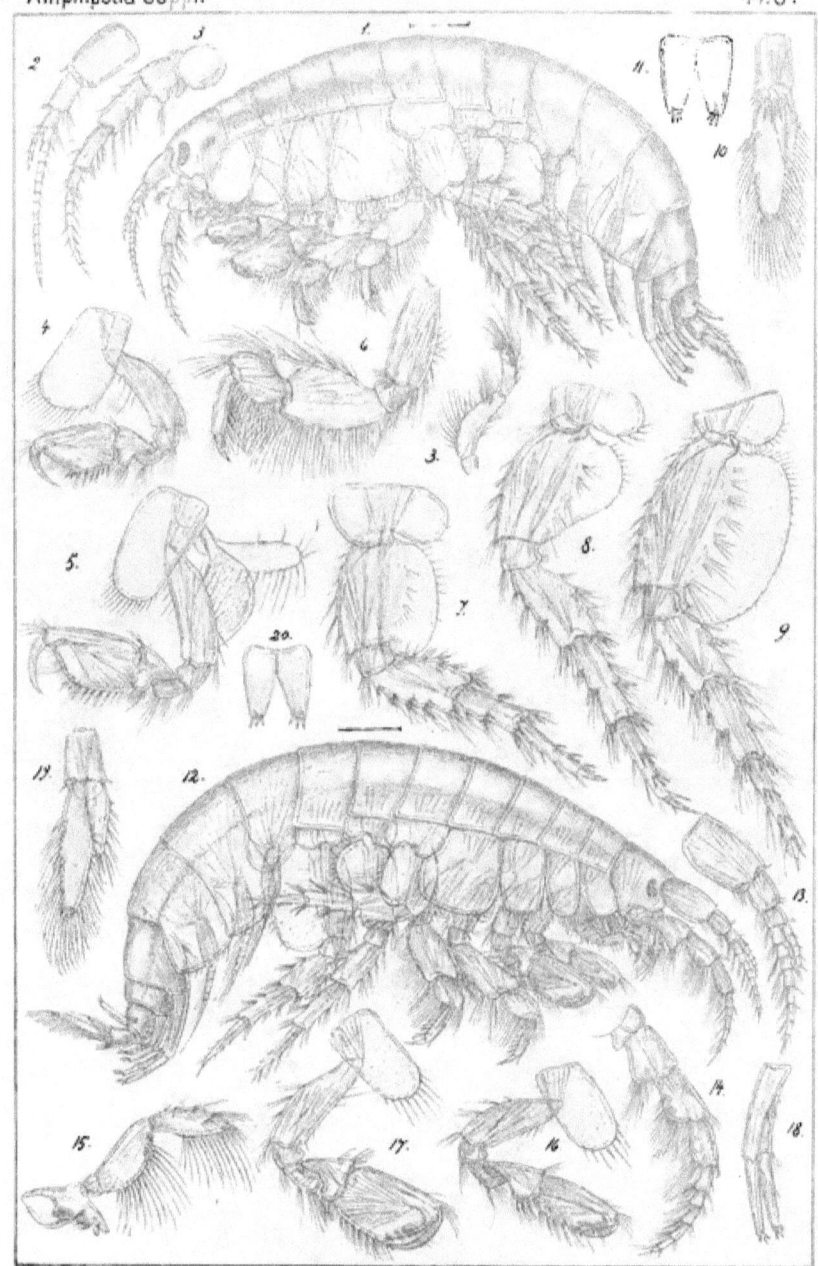

G.O.Sars autogr.   Gammarus Weidemanni, G.O.Sars.
mæosticus, Sowinsky.

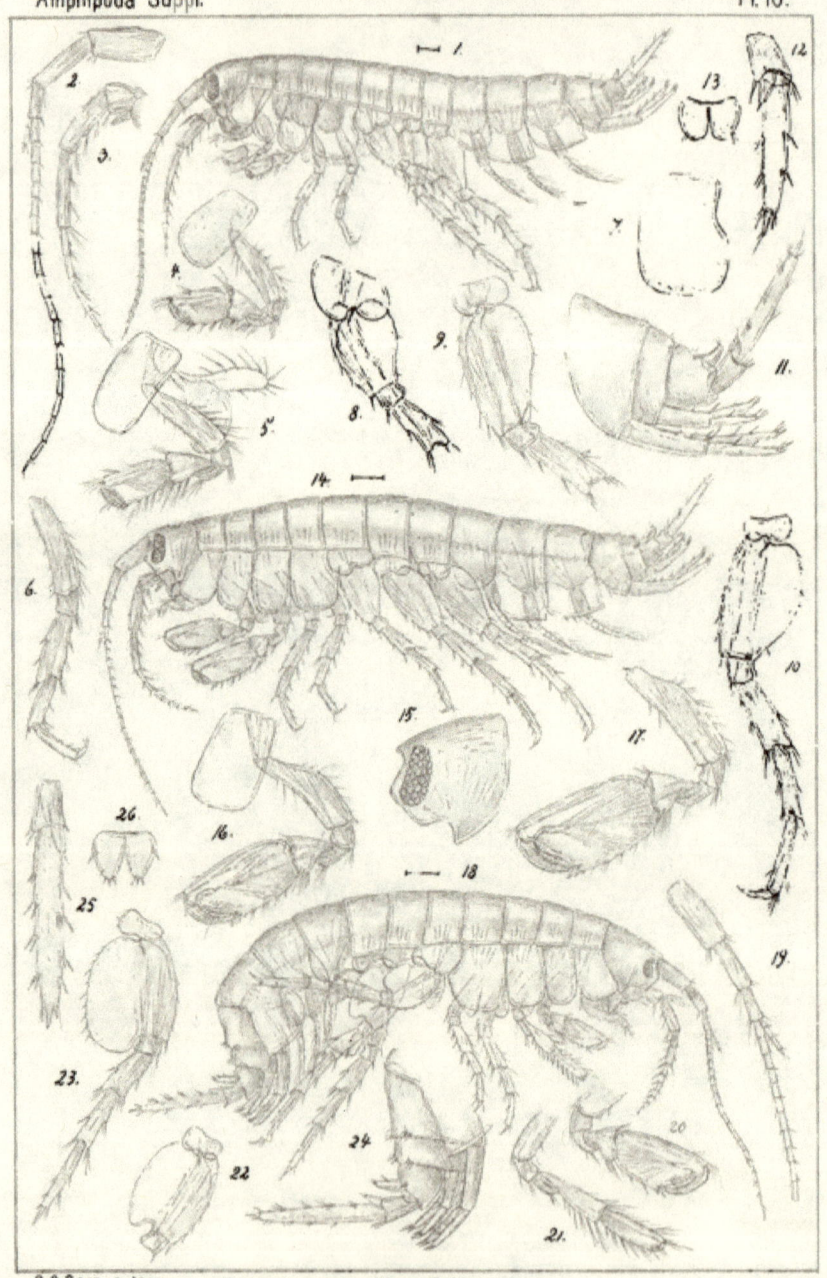

Gammarus pauxillus, Grimm.
" Andrussowi, G.O.Sars.

Amphipoda Suppl. G.O.Sars Crustacea caspia. Pl. II.

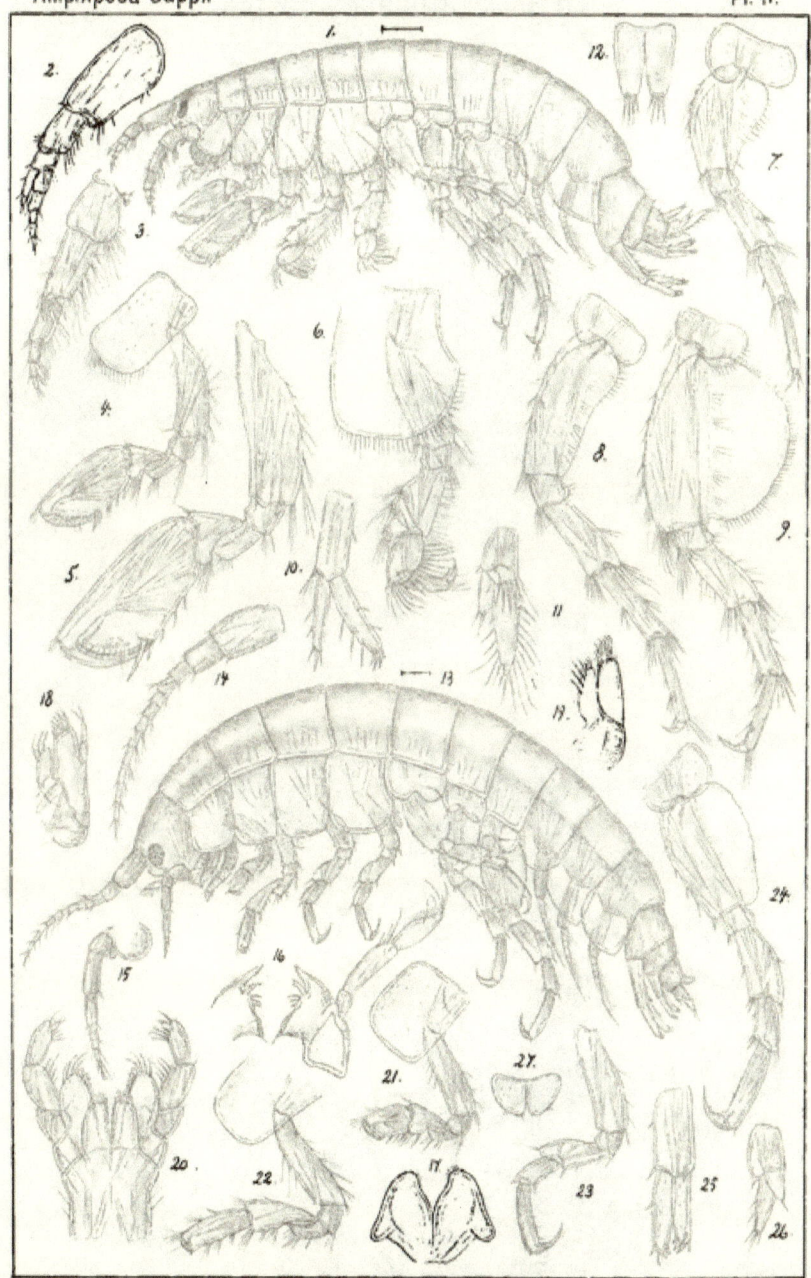

G.O.Sars autogr. Niphargoides Grimmi, G.O.Sars.
Cardiophilus Baeri, G.O.Sars.

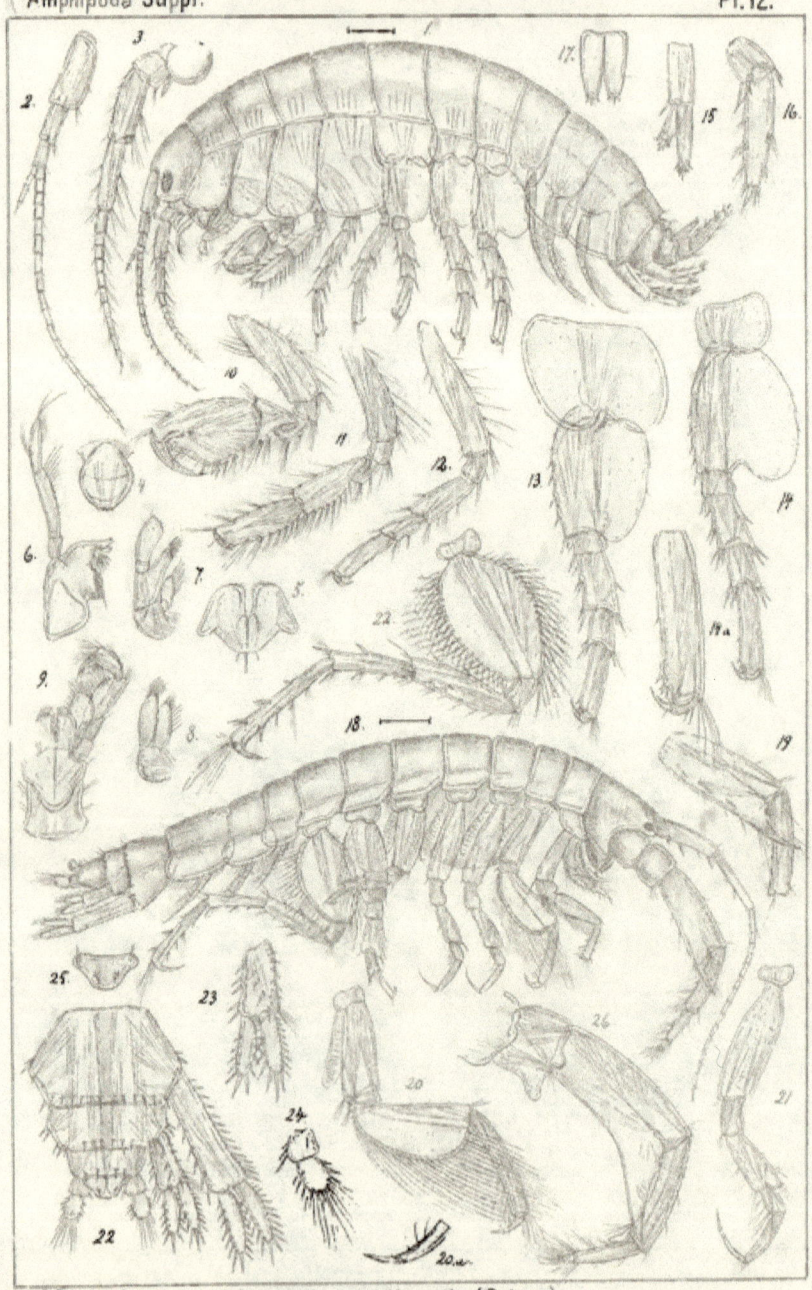

Iphigenella acanthopoda, (Grimm).
Corophium spinulosum, G.O.Sars.

www.ingramcontent.com/pod-product-compliance
Lightning Source LLC
Chambersburg PA
CBHW032007230426
43672CB00010B/2274